Vortex Publishing LLC.
4101 Tates Creek Centre Dr
Suite 150- PMB 286
Lexington, KY 40517

www.vortextheory.com

© Copyright 2019 Vortex Publishing

All rights reserved. No part of this book may be reproduced or transmitted in any form or by any means, electronic or mechanical, including photocopying, recording or by any information storage and retrievable system without the prior written permission by the Publisher. For permission requests, contact the publisher.

Printed in the United States of America

1 2 3 4 5 6 7 8 9 10

Library of Congress Control Number: 2019953542

ISBN   978-1-7332996-6-4
eISBN 978-1-7332996-7-1

Editor's note: All drawings in this book are original illustrations made by Dr. Moon. They are kept as they are to maintain the integrity of his work.

TABLE OF CONTENTS

OVERVIEW IV
AUTHOR'S NOTE V
From me to you… V

PART I
ACCORDING TO CONTEMPORARY SCIENCE …

Chapter 1: The Problem With 20th Century Science's Creation of the Universe Theory 1
Chapter 2: Contemporary Science's Explanation for the Creation of the Universe and Its
      Major Problems! 3
Chapter 3: Contemporary Science's First Five "Early Epochs" 5
  (A.) the Planck epoch;
  (B.) the grand unification epoch;
  (C.) the inflationary epoch;
  (D.) the electroweak epoch;
  (E.) the electroweak symmetry breaking epoch;
Chapter 4: Contemporary Science's Second Five: "Early Epochs" 7
  (F.) the quark epoch;
  (G.) the hadron epoch;
  (H.) the lepton epoch;
  (I.) the photon epoch;
  (J.) the recombination era

PART II
ACCORDING TO THE VORTEX THEORY OF ATOMIC PARTICLES…

Chapter 5: Introduction of the Creation of the Universe Using The
      Vortex Theory of Atomic Particles 9

PART III
THE EVIDENCE!

Chapter 6: Time Does Not Exist; Space Is Made of Something; There Is No Higgs Field, Or
      Higgs Boson!!! 11
Chapter 7: Protons and Electrons Are Holes in Space Connected by Tiny
      Higher Dimensional Vortices 14
Chapter 8: Space Is Not a Void: It Can Bend, Stretch, and Flow 17
Chapter 9: How Subatomic Particles With a Charge of Zero Are Formed 21
Chapter 10: How the "FIVE" Forces of Nature Are Created 25
  The Force of Gravity
  The Electromagnetic Force
  The Weak Force
  The Strong Force
  The Anti-gravity Force
Chapter 11: The Reason Why Neutrinos Spin CCW and Anti-neutrinos Spin CW Reveals That
      There Are Two Sides of Space 35

Chapter 12: There Are Two Volumes of Space, One in Expansion One in Contraction; All Created by One Massive Particle Turning Inside Out..37
Chapter 13: Each Volume Contains at Least 7 Dimensions 42
Chapter 14: The Beach Ball Analogy 44
Chapter 15: Implications of Space Turning Inside Out; & the Ant and the Basketball! 45
Chapter 16: The Three-Dimensional Space We Live In Is an Illusion...46

PART IV
THE CREATION AND DESTRUCTION OF THE UNIVERSE!

Chapter 17: FINALLY: The Creation and Destruction of the Universe! 47
  *The oscillating Universe
  *Dynamics of the Particle
  *Subatomic Vortices Reveal the Existence of 4D Space
  *Everything in the Universe is Created out of Space
  *The Illusion
  *Space Turns "Inside out": The Beach Ball Analogy…
  *Example of Space turning Inside out: the Neutron
  *The Destruction of the universe: in the Blink of an Eye!!!
  *But it will Instantly Begin Again
  *The "Re-creation" of Matter
  *Creation of the CMB Background Radiation
  *Creation of Galaxies and Regions of Dark Matter
  *Abnormalities in the Incoming Volume of Space from the Former Contracting
    Side Creates Regions where Galaxies form
  *The End of this Cycle

PART V
SOME INTERESTING ILLUSTRATION?

Chapter 18: Analogous Drawings Used to Illustrate the Beginning of the Universe 52
Chapter 19: Analogous Drawings to Illustrate How the Particles of Matter Are Created............57
Chapter 20: What Ended the Initial Creation? 61
Chapter 21: The Shape of the Universe Is the Solution to Seeing CMB Radiation 66
Chapter 22: A Problem With the Red Shift! 69
Chapter 23: The Secret of the Universe 70
Chapter 24: An Experiment for Astronomers & Astrophysicists! 71

Another Nobel Prize winning Experiment is proposed, this time for Astronomers & Astrophysicists! Because the Vortex Theory of Subatomic Particles reveals that the three dimensional volume of space of the universe is really the surface of higher dimensional space, we are looking into a universe that is an illusion! This illusion can be revealed by a revolutionary experiment that allows astronomers and astrophysicists to look into a universe that is an illusion

Chapter 25: Determining the True Age of the Universe? 75
Conclusion..76
References **79**

PART VI
THREE EXPERIMENTS PROVING THE VORTEX THEORY OF ATOMIC PARTICLES IS TRUE!

The Ball and Zinc Experiment…The Proof of "The Vortex Theory" 87
The St. Petersburg State University Experiment That Discovered The
"Photon Acceleration Effect" 108
The Discovery of the Fifth Force in Nature: The Anti-gravity Force 116

OVERVIEW

Originally meant to be only a three part trilogy called *The End of the Concept of Time, Parts I, II, and III,* it was soon realized that all of the information presented in the first three books was incomplete without being able to use it to explain the creation and eventual destruction of the physical universe.

Unlike 20th Century science's explanation for the creation of the universe via the Singularity, that has no proof, but is based totally upon supposition, presented here is the shocking proof for the real vision for the construction of our universe - a fantastic vision unlike anything anyone has ever imagined before! A vision that reveals our entire universe is really one gigantic multi-dimensional "particle!" What we call "space" is not a vast void at all; but is actually the substance this particle is made of!

According to the evidence discovered about the creation of matter, space, energy, time, and the forces of nature listed in Books 1, 2, & 3, what we call the three dimensional physical universe appears to have been created by one single gigantic multi-dimensional particle turning inside out; creating two separate massive multi-dimensional volumes, one in rapid expansion, the other in rapid contraction; with our three-dimensional space their mutual surface sandwiched between them.

The subatomic particles everything in the universe is made of are not little solid particles as is presently believed, instead, they are really three-dimensional holes that the three dimensional space from the expanding volume flows into and out of; creating tiny vortices of whirling space extending into the contracting side's fourth dimensional volume. Although from our point of view, all matter appears to be separate from all other matter, this is only an illusion. Every infinitesimally small subatomic "particle" in the universe that atoms are constructed out of are really part of the one massive gigantic particle: including the tiny subatomic particles we are made of. We are all part of it; nothing is separate from anything else!"

But will it last forever?

Just as the beginning of the inside out twist in space started an instant expansion of one volume out the other in a single instant of time, creating what science calls the "Big Bang", the end will also occur in a single instant of time: in the blink of an eye!

When all of the space from the contracting volume has flowed into the expanding volume, the cycle will be complete; all the vortices will instantly disappear; in a single instant, all of the matter of the universe will instantly disappear, everything will be gone. Including us!

But even more amazing, just after this instantaneous universal destruction ends, it will instantly begin all over again! Just like it has happened many times before in the past, the instant the contracting volume disappears, a rip in space will instantly reoccur in the former expanding volume's geometric center, and another "Big Bang" will create another new universe. Throughout the rest of this book, we will present the evidence of this shocking thesis.

AUTHOR'S NOTE:

This first ever book uses pictures to explain what was previously one of the most difficult subjects in all of physics. But no more! This revolutionary breakthrough in science now makes it possible for High School Students to understand what the greatest scientists of the 20th Century were unable to comprehend. Although some minor math is necessary, it only takes an eight grade level to understand it.

Also, even though some of the subatomic particles of nature are talked about in this book, making it seem as if a special education is necessary to understand them, this is not so. Later on, all of the terms in every chapter will be so easily explained, that every-one can understand them. And it is very important that everyone does understand them: because these infinitesimal particles are the key to understanding our ultimate fate! So do not let a misconception that you have to be a rocket scientist cheat you from learning the most fantastic discovery ever made about yourself and your universe.

From me to you...

To all you good friends and to all you good friends I will never meet, welcome to the explanation for the creation of the universe. This explanation was made possible by the use of the many scientific discoveries of the past 25 years that were made using the *Vortex Theory of Atomic Particles*. This addendum to the first three books in this series, was made because many hypothesis made by past and contemporary scientists about the creation, shape, and destiny of the universe are simply not true.

To correct these mistakes, in Chapter 24 a revolutionary experiment is proposed for Astronomers and Astrophysicists. This experiment will allow these scientists to discover how the universe was really created and see its true shape: a shocking vision unlike anything anyone has ever imagined before!!!

In honor of two of Russia's greatest scientists: Dr., Prof. Konstantin Gridnev; and Dr., Prof. Victor V. Vasiliev, I have given the vortices the name "Konsiliev Vortices".

Sincerely,

Russell Moon

PART I
ACCORDING TO CONTEMPORARY SCIENCE...

Chapter 1
The Problem With 20th Century Science's
Creation of the Universe Theory

> According to 20th Century science, the mystery of the creation of the universe is believed to be solved with the ideas of the "Big Bang" and what is called the "Singularity". But on second glance, serious problems are revealed in these two theories that scientists either ignore or fail to acknowledge...

We have all heard of the Big Bang, the giant explosion that began the universe and set matter into motion; and of course the "thing" that exploded: the "Singularity", the micro-sized little dot that contained all of the matter of the universe within it. However, a little research into the history of the Big Bang now reveals a big problem! This theory was proposed almost a hundred years ago before the hundred billion plus galaxies were discovered that are now known to populate the universe. So how can all the matter in a hundred billion plus galaxies be compressed into a single dot no bigger than a proton?

The first individual who came up with this theory was not Edwin Hubble as the history books proclaim; instead, it was Georges Henri Joseph Édouard Lemaître: a Belgian Catholic Priest, astronomer and professor of physics at the Catholic University of Leuven. Lemaître published his theory in 1925, two years before Hubble published his article in 1927. In 1925, Lemaître proposed the Big Bang, the Singularity [also called the "Cosmic Egg"], and the expanding universe! But if he knew there were a hundred billion galaxies, would he still try to theorize that all of the matter – in all of these galaxies – was compressed into one single dot?

Today, he would also be faced with the problem of Black Holes. When matter is compressed into a Black Hole, its volume grows larger not smaller. If Lemaître was right about the Singularity, Black Holes should grow smaller and smaller instead of larger and larger as matter is compressed into them. And then there is the problem with the point of origin of the Big Bang?

The theory of the Big Bang creating the universe is unequivocally true say scientists. They say that the discovery of the "Background Radiation" reveals that the Big Bang did indeed happen. Also, the "red shift" of the light coming from distant galaxies reveals that the universe is expanding outward as a result of the Big Bang. All true...until you ask this one little question, "Where is the Point of Origin of the Big Bang?" "Or more specifically, where is the point in space where the Singularity exploded and began its expansion?"

As was mentioned in Book 2 of this series, in 1054 AD, the people of the earth witnessed a gigantic supernova. Although they did not know what they were seeing, the Chinese wrote about a new star that suddenly appeared in the sky and was so bright that it could be seen in the middle of the day. In the southwest, Native Americans drew stars on rocks in awe of this one of a kind event. Today, in our telescopes, we can witness the remnants of that explosion in the Crab Nebula.

By looking at the Crab Nebula and measuring the Doppler shift of its light, we can tell by the velocity of the particles streaming out into the universe that the nebula is the result of a gigantic supernova explosion that occurred approximately 1000 years ago [approximately equaling the year 1054 AD]. We can also use the same technique to trace the coordinates of the particles streaming out into the universe from the Crab Nebula to determine the point of origin of this supernova. We can see approximately where the blast occurred and witness the now existence of a small rapidly spinning neutron star that is the left over remnant of this once massive star.

So, how come we cannot do this with the Galaxies of the universe? How come we cannot use this same technique to determine the point of origin of the Galaxies created after the Big Bang?

Unfortunately, this same scientific method fails when we try to apply it to the motion of the Galaxies out of which the entire physical universe is constructed. Even more amazing, we find it hard to believe that when we look for the origin of the Big Bang, we discover *that **every point in the physical universe** appears to be the point of origin we are looking for! How can this be?* [This observation will become extremely important later on when we discover the true shape of the universe.] Because if indeed the Singularity existed, the point of its explosion outward would, like the Crab Nebula, have one single point of origin and not an infinite number of points!

Many years ago, it was this failure to explain the Point of Origin that made me realize something is wrong, that there appeared to be a problem with this simplistic explanation for the creation of the universe every professor of astronomy in college was teaching, and every student was blindly accepting without question!

So how can this problem be resolved?

At the time I was introduced to this problem, I was just a freshman student without the knowledge necessary to offer a solution. However, all that is now changed.

A strange set of circumstances forced me to re-investigate what has come to be called the Quark Theory. Amazingly, found in the knowledge explaining the tiniest of all the pieces of the universe is also found the explanation of the most massive aspects of it!

To revolutionize the explanation for the creation of the universe, we first present Contemporary Science's vision of the universe's creation using the hypothesis of the Singularity and the Big Bang along with the inherent problems of each. Then we will give the explanation discovered using The Vortex Theory of Atomic Particles. Nothing will ever be the same again!

Chapter 2
Contemporary Science's Explanation for the Creation of the Universe and Its Major Problems!

Contemporary Sciences explanation for the creation of the universe possesses big problems that are never talked about. In this chapter we will discuss them and reveal knowledge that the general public has been kept unaware of.

THE SINGULARITY?

According to contemporary science, everything that exists today in the universe supposedly began with the "Singularity". This tiny dot smaller than a single proton allegedly had all of the matter of over a hundred billion galaxies in the universe compressed inside of it!

The singularity ⟶ .

Then, the dot somehow blew up in a massive explosion called the "Big Bang".

Figure 2.1

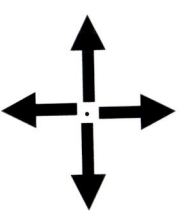

Problem with this idea…

Nobody in the world can explain where the singularity came from! Did it just pop into existence? If not, how long was it there? And what created the explosion!

Furthermore, where is the point of origin of this massive explosion? Again, all explosions such as the huge supernova blast in 1054 AD, seen in the daytime sky and recorded by the Chinese have a "Point of Origin". So where is the point of origin of the Big Bang? All of the galaxies in the universe should be streaming away from this single point!

Contemporary science explains that this dot of matter raced outward for hundreds of thousands of years creating all of the protons and electrons that exist in the universe today. However, nobody can explain what happened to all of the anti-protons and positrons!

According to the discovery of the phenomenon called "Pair Production" [verified time and time again in linear accelerator collisions] there has to be one anti-proton for every proton created; and one positron (the anti-particle of the electron) for every electron created. So, what happened to all of this anti-matter? [Nobody knows?] Some say that the anti-matter was annihilated during this epoch. But this is flawed logic: it takes a proton to annihilate an anti-proton and an electron to annihilate a positron, so if this annihilation really occurred, there would be no protons or electrons left in the universe!

Chapter 3
Contemporary Science's First Five "Early Epochs"

> Contemporary science has divided the early universe into ten major epochs. Each one of which has many mistakes in it: (A.) the Planck epoch; (B.) the Grand unification epoch; (C.) the Inflationary epoch; (D.) the Electroweak epoch; (E.) the Electroweak symmetry breaking epoch; (F.) the Quark epoch; (G.) the Hadron epoch; (H.) the Lepton epoch; (I.) the Photon epoch; (J.) the Recombination era; etc. The first five will be discussed in this Chapter.

THE FIRST FIVE EPOCHS BEGIN WITH THE PLANCK EPOCH…

(A.) In the "Planck Epoch", the temperature of the universe was assumed to be so great that the four [known] forces of nature - gravity, weak force, strong force, and electromagnetic force - were one force.

> However, one of the discoveries of the Vortex Theory is that the four forces of the universe [plus a fifth anti-gravity force] are not different manifestations of "one force"; but rather, are created by bent and flowing space [while the strong force is created by a constant exchange of quarks between protons and neutrons in the nucleus of an atom]. Hence, the assumption of the Planck Epoch that all four [known] forces were originally merged into one force is wrong.

(B.) Here, in the "Grand unification Epoch" the three "gauge interactions" [particles] of the Standard Model which define the electromagnetic, weak, and strong forces are merged into one single force.

> This again is a mistake. Just as explained above, neither the four forces, nor three of them, cannot be unified either by force or boson particles [particles believed to transfer force]. Each is uniquely created by space itself as will be explained later.

(C.) The next epoch called the "Inflationary Epoch" is a strange period of time where the expansion was accelerated. It was subsequently theorized that this epoch could be proven by the detection of "inflationary" gravitational waves. These were supposedly discovered in 2014 by the BICEP2 group. However, in 2015, this was proven wrong. Instead, the cause of the findings of the BICEP2 group were recognized as being the result of polarized dust in the Milky Way Galaxy and not gravitational waves.

> At the present moment there is no known explanation in contemporary science for the acceleration of matter: it is only speculation! [Later we will explain how it happens.]

(D.) In the <u>Electroweak Epoch</u>, the temperature of the universe dropped allowing the strong force to separate from the weak force.

> The problem with this postulate is the following: according to the discovery of the Vortex Theory of Atomic Particles, as explained before [and will be explained later], these two forces are created so differently they could never be unified. Hence such an epoch could never have happened.

(E.) In the <u>Electroweak Symmetry Breaking Epoch</u>, contemporary science believes that the "Higgs field" spontaneously acquires a "vacuum expectation value" (what is this?) breaking "The electroweak gauge symmetry" (again, what is this?). This causes the gauge bosons [the W and Z bosons] of the electromagnetic and weak forces to manifest differently. Also because of the "Higgs mechanism" (and again, what exactly is this?), all elementary particles interacting with the Higgs field become massive after being mass-less at higher energy levels (this is utter foolishness: something mass-less becoming massive)! All these weird sounding theories sound impressive, but are nothing but utter foolishness!

<u>Shockingly, all of the above is total fiction! One of the great discoveries of the Vortex Theory reveals that there is no Higgs field or Higgs boson! [Someone at CERN must have made a mistake! A scientific study shows the possibility exists as will be revealed in this book!]</u>

Chapter 4
Contemporary Science's Second Five: "Early Epochs"

> Contemporary science's last five Epochs are given here. Again, each one has major mistakes in it: (F.) the Quark epoch; (G.) the Hadron epoch; (H.) the Lepton epoch; (I.) the Photon epoch; (J.) the Recombination era.

THE SECOND FIVE EPOCHS BEGIN WITH THE QUARK EPOCH…

(F.) In the Quark Epoch, the universe was filled with a dense plasma made out of quarks, gluons, leptons [electrons], and their anti-particles.

> There is one big problem with this postulate, the leptons and their anti-particles most likely existed; but the Vortex Theory discovered that quarks are higher dimensional holes existing within the three dimensional holes we call protons and neutrons. If gluons exist, they are also higher dimensional creations made out of bent and flowing space. Subsequently, they cannot exist within three dimensional space because fourth dimensional creations [especially holes] cannot exist outside of their surfaces! [For example: the interior three dimensional volume of a basketball cannot exist outside its two dimensional surface.] Since quarks exist in higher dimensional space, [and cannot exist outside of higher dimensional space], this is the reason why scientists have never seen a "naked quark" [a quark existing in three dimensional space outside a particle such as a proton].

(G.) In the Hadron Epoch, the quark gluon plasma cloud cools down allowing protons and neutrons to form: [note: a hadron is a particle containing quarks.]

> Again, this never happened because quarks [higher dimensional holes] cannot exist outside of a "particle" [a three dimensional hole] on their own. Hence, there never were free quarks and gluons packed together in a plasma cloud.

(H.) The <u>Lepton Epoch</u> is a strange epoch. Here, most of the hadrons, anti-hadrons; and leptons, anti-leptons annihilate each other leaving only a small amount of leptons.

> This epoch is also suspect because the Hadron epoch would never have happened. This epoch assumes the Hadron epoch with all of its quarks and gluons existed. Because epoch (G) above never happened, this epoch could not have happened either!

(I.) According to contemporary science, the next epoch, the <u>Photon Epoch</u> lasted for about 380,000 years after the big bang. Here, photons were trapped inside and interacted with a massive cloud of charged particles.

> This appears to be closer to the truth. Nothing can be found to dispute this postulate, except for the fact that if the above epochs that preceded it never happened; and if so, then the time-line for this one is wrong too!

(J.) The <u>Recombination Era</u> is important. Contemporary science postulated that this era began the creation of hydrogen atoms; allowing the photons that were "randomly walking" [bouncing around, trapped inside the dense plasma cloud], to "decouple" [to leave] and finally "free stream" out into the universe that at last becomes transparent. These photons that no longer interacted with matter constitute what is today called the cosmic microwave background radiation: the CMB.

> At present, despite the apparent disparage with the time-line as just discussed in the photon epoch, the Recombination Era becomes extremely important. It is important because this CMB background radiation streamed out into the universe <u>before</u> the creation of the galaxies. As such, these photons should have streamed past our galaxy [that formed much later after the photons left], and traveled out into the space of the universe, never to be seen again!
>
> However, because we can now see this CMB radiation, it allows us to arrive at a most astounding conclusion about the physical shape of the universe! But first, before we present this fantastic discovery, we must begin with the Vortex Theory of Atomic Particles explanation of how the universe was created.

PART II
ACCORDING TO THE VORTEX THEORY OF ATOMIC PARTICLES...

Chapter 5
Introduction of the Creation of the Universe Using the Vortex Theory of Atomic Particles

According to the discoveries presented in Books 1, 2, & 3, about the Vortex Theory of Subatomic Particles, and 40 scientific papers presented in international scientific conferences throughout the world over the past 20 years, the entire universe is revealed to be one single massive multi-dimensional particle in expansion. But how did it get this way? How did it begin and how will it end?

Using the scientific discoveries made by the Vortex Theory of Atomic Particles, the explanation for the creation of the universe presents an absolutely shocking vision unlike anything anyone has ever envisioned before. Although there are several ways to present this vision and its proof, each has its merits and drawbacks. Therefore, after considering all the possibilities, it was finally decided to present this knowledge in the following manner: present a brief synopsis of the theory first, next present the evidence, then finally put it all together in what has become a shocking exposé, that again, is unlike anything anyone has ever seen before. So we begin with presenting 9 important principles of this theory that will be investigated and expanded in the body of this book…

THE BRIEF SYNOPSIS: PRESENTED IN 9 IMPORTANT PRINCIPLES

- ❖ We live in an oscillating universe: one that appears to have been created and destroyed many times in the past.
- ❖ We cannot say when the sequence began or when it will eventually end, only how this particular one began, how it will end, and how it will start all over again: if in fact it does?
- ❖ The whole of the universe is one gigantic multi-dimensional "particle" that is turning inside out.
- ❖ This gigantic "particle" possesses at least seven dimensions. There are possibly more but only later discoveries in particle physics will determine how many.
- ❖ The shape of this "particle" can be roughly described from our limited perspective as a gigantic multi-dimensional sphere.
- ❖ The three dimensional volume of space we live in is an illusion. We live upon the three dimensional space that is the surface of the higher dimensional volumes of space this three dimensional surface encapsulates.
- ❖ This gigantic "particle" is turning inside out; creating for all practical purposes two volumes of space: one in expansion, one in contraction.

- ❖ The three dimensional surface we live upon is in motion. It is expanding outward growing larger and larger as the volume of the contracting particle flows into the volume of the expanding volume.
- ❖ When all of the contracting volume has flown into the expanding volume, all of the matter of the universe will instantly be destroyed; it will all disappear in the blink of an eye!

However, it will instantly start all over again, creating a brand new cycle!

In PART III coming up next, evidence will be presented for all of the above and for all of the ramifications that follow.

PART III
THE EVIDENCE!

Starting with this chapter and continuing through the rest of **PART III**, the evidence for the creation of the universe via the Vortex Theory of Atomic Particles is presented. The sentences in red at the beginning of each chapter reveal important point or points of each chapter that later, will all be tied together in the explanation for the creation of the universe in PART IV along with those 9 previously stated in Ch 5. All of these individual pieces of evidence are critical for both the explanation of the creation and the eventual destruction of the universe.

At the end of each chapter in this section, references are given. Note: the 20th Century contemporary vision for the creation of the universe using the singularity has no references! It is based purely upon speculation; upon the hypothetical existence of a microscopic particle called the Singularity with absolutely no evidence to support its existence!!!

Chapter 6
Time Does Not Exist; Space Is Made of Something;
There Is No Higgs Field, Or Higgs Boson!!!

Time does not exist; hence, there is no fourth dimension of "Space-time". Space is made of something possessing unique dynamics; there is no Higgs field! Gravity is created by less dense space! This discovery was made in 2005; a PhD was awarded for its thesis and proof.

This discovery allows us to subsequently make our first most important discovery about the creation of the universe: that there is a fourth dimension but it is a pure dimension of space and contains no time characteristics. The current belief that space is a void is a mistake. Space is made of something that can bend, stretch, and flow; that can also become denser or less denser. Later, the true explanation of the force of gravity and the creation of mass will reveal there is no Higgs Boson or Higgs field.

Many years ago, the discovery of the Vortex Theory of Atomic Particles revealed that time does not exist and space is made of something. In 2005, a PhD [from the Russian Ministry of Education; number KT#032771] was awarded for this discovery and the mathematical analysis that proved it. The thesis was published by the St. Petersburg Branch of the Russian Academy of Sciences in 2012, and the startling conclusion of this shocking paper is presented here. It is both the beginning and foundation for everything that is to follow.

Although it is hard to believe, the tiniest particles of the universe are the key to understanding its most massive parts. Since the key to understanding how matter is formed was ironically discovered with the fact that time, "time" itself does not exist, we begin the proof of the creation and the destruction of the universe with this most shocking discovery.

Conclusion from the *"End of Time"* thesis:

The era of the Theory of Relativity comes to an end when it is realized that Albert Einstein's vision of the universe is based upon the effects we see and not upon the cause of these effects.

From Einstein's point of view, the motions of everything in the universe were "relative" to the motions of everything else. This relationship is not the true reality; nevertheless, it is a real effect. Hence, the Lorentz transformation equations that allow us to calculate the "time differences" between two moving frames of reference are still valid. In addition, even though the "twin's paradox", the orbit of Mercury aberrations, and many other observations from the "relativistic" vision of the universe are still real effects, the causes of these effects have nothing to do with a fourth dimension of "space-time".

Although it is true that the space surrounding the sun and other large interstellar objects appears to be bent, it is not the "bend" that is responsible for the creation of the force of gravity. The "bend" is a phenomenon created by the less dense region of space surrounding the sun. This less dense region of space is responsible for the creation of the force of gravity and for the bending of starlight seen during eclipses.

Although many ideas regarding time and space have to be discarded or amended, many mathematical formulas are still usable. Just as Newton's laws are still applicable even though the Theory of Relativity amended them, the Theory of Relativity is still applicable even though the Vortex Theory amends it; because the Theory of Relativity is a real phenomenon.

However, in the "microscopic" world inside the atom, things are very different. The view of the universe from inside a proton or an electron is completely different from the relativistic view. Completely different too is our vision of time.

According to relativity, time exists along with matter, space, energy and the forces of nature as one of the five fundamental "pieces" of the universe. Furthermore, according to Albert Einstein, time exists as part of a fourth dimension of the universe called "space-time". However, this idea is and always was conjecture. No fourth dimension of time has ever been discovered.

Also, according to the relativistic vision of the universe, time flowed at the speed of light: the maximum speed possible in the universe. Consequently, because any velocity "V" of matter cannot add to the speed of time, the velocity "V" of matter creates time dilation effects. But this too is a mistake. According to the discoveries in this paper, we can now demonstrate that it is the tiny vortices themselves flowing at the speed of light between particles that are responsible for creating the phenomenon of time; and we have seen that at near light velocities, it is the slower *apparent* velocity of the vortices that are responsible for creating the phenomenon of time dilation.

Although the idea that "time does not exist" seems offensive to some, this unpleasant emotion can be dispelled by understanding that modern peoples did not invent time. The concept of time is so old it predates writing. It appears to be invented by ancient peoples during their efforts to keep track of the seasons in order to plant crops. But no matter what the reason, the fact still remains that "time" was invented by people totally ignorant of the construction of the physical universe. People who did not even know the world was round!

Another important misunderstanding that must be cleared up regards clocks. Although some people believe that clocks keep track of time, this too is a mistake. Clocks are only associated with time. Clocks keep track of a position on the rotating surface of the earth in relation to the Sun when it is directly overhead.

The length of an hour, minute, and second is arbitrary. They are only useful upon this planet. When men go to Mars, earthly clocks will no longer allow the user to predict when the sun will be directly overhead. Because Mars rotates approximately one-half hour longer than the earth, in twenty-four Martian days, when the earthly clock indicates it is noon, it will really be midnight on Mars. Hence, the length of the hour, minute, and second will have to be modified or scrapped altogether on Mars.

Excerpt about Space being made of something from the thesis:

SPACE:

According to the Vortex Theory of Atomic Particles, space is made of something. This is *not* a return to the Aether Theory where it was mistakenly believed space was made of something and matter was made out of condensations of space: like ice in water. Because as matter moves, such a scenario creates an Aether wind: and the Aether wind was eliminated by the Michelson Morley Experiment.

MATTER:

According to the Vortex Theory, "particles" of matter such as protons and electrons are not particles at all. Instead, they are discovered to be three-dimensional (3d) holes existing upon the surface of fourth dimensional (4d) space. As these holes move through 3d space, 3d space reconfigures around them and no Aether wind is created.

Also from Book 3…

THERE IS NO HIGGS FIELD!!! Later, because the phenomenon of mass and the force of gravity will be shown to be created by less dense space, it is revealed that there is no Higgs field or Higgs boson! [How did CERN state they discovered the Higgs boson particle when it is impossible for it to exist?]

See reference number 8. on page 84

Chapter 7
Protons and Electrons Are Holes in Space Connected by Tiny Higher Dimensional Vortices

> The sub-atomic "particles" of matter that all atoms are created out of are really three dimensional holes extending downward into a fourth dimensional volume of space. The particles of matter and anti-matter are connected by tiny three dimensional vortices of spinning space that extend into the fourth dimensional volume of the contracting side.
>
> This initial discovery allowed us to first understand that our universe possesses at least one more dimension of space; and that the three dimensional space we can see is one piece, one single volume: one gigantic particle of simply enormous size!

Again, the following excerpt from the *End of Time* thesis, revealed the discovery that sub-atomic particles of matter are really three dimensional holes:

MATTER

According to the Vortex Theory, "particles" of matter such as protons and electrons are not particles at all. Instead, they are discovered to be three-dimensional (3d) holes existing upon the surface of fourth dimensional (4d) space. As these holes move through 3d space, 3d space reconfigures around them and <u>no</u> Aether wind is created.

The proton is a 3d hole bent into the surface of 4d space; 3d space flows into the proton creating its electrostatic charge. The electron is a 3d hole bent out of 4d space; 3d space flows out of the electron creating its electrostatic charge.

Figure 7.1

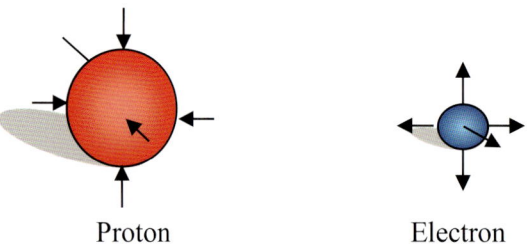

Proton　　　　　Electron

Because space is pulled into the proton it is surrounded by a region of less dense space. Likewise, because space flows out of the electron, it is surrounded by a region of denser space. These import concepts will be elaborated upon later.

[Note: although the quarks within protons are also explained by the Vortex Theory, it was subsequently discovered that the 4d vortex does not flow through them; hence, they are not necessary for this mathematical proof.]

Furthermore, it is hypothesized that these holes are really the ends of invisible 4d vortices existing in 4d space, connecting the proton to an anti-proton, and the electron to a positron. Three dimensional space flows into the proton, through a 4d vortex and exits out of the anti-proton; likewise, 3d space flows into the positron, through a 4d vortex and out of the electron.

The proton and the electron are connected by an invisible vortex of 3d space flowing from the proton to the electron in 4d space.

Figure 7.2

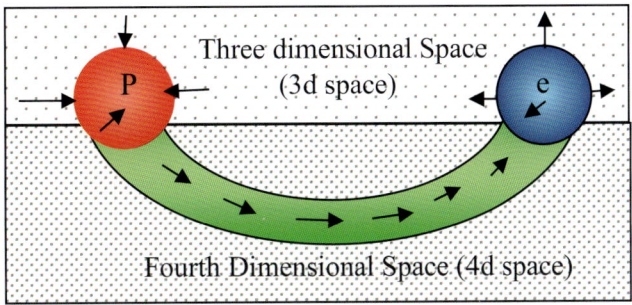

WHEN A HYDROGEN ATOM IS CREATED

THE TWO VORTICES:

When a proton captures an electron creating a hydrogen atom, the vortices that connect them to the anti-proton and the positron break, reconnecting the proton to the electron, and the anti-proton to the positron.

If the proton and the electron are pulled closer to each other by their electrostatic forces, a second vortex is created in 3d space when all of the space flowing out of the electron begins to flow into the proton: creating a situation seen below in Figure 7.3. Space now flows into the proton, into 4d space through the 4d vortex, back into the electron; then out of the electron and through 3d space and back into the proton: creating a circulating flow containing a fixed volume of space.

Figure 7.3

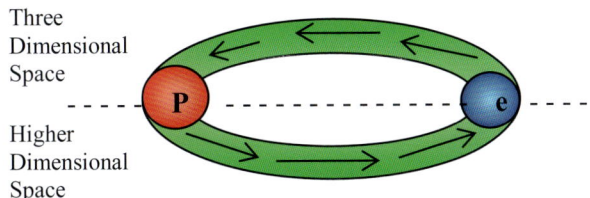

THE HYDROGEN ATOM:

Because the hydrogen atom is the simplest of all atoms, it is the example used in this analysis. However, since all protons and electrons *in* all atoms everywhere would also be connected by two vortices of flowing space, the principles introduced here apply to all other atoms in the universe as well.

When the circulating flow commences, both electrostatic charges are neutralized. The word "neutralized" was used because no flowing space escapes from the system. [Note: ions are created when two atoms in a molecule are separated and a proton in one atom is connected to an electron in another atom via a 4d vortex that still flows between them.]

ADDENDUM:

Because there are no lines of demarcation in the volume of the three dimensional space in the universe, we can conclude that the entire volume of three dimensional space we can see is one volume: one single piece!

See references number 8. on page 84 and number [26] on page 81

Chapter 8
Space Is Not a Void: It Can Bend, Stretch, and Flow

> Other particles identified as photons of energy are really condensed packets of three dimensional space thrown from the vortices. Below is explained the particle and wave aspects of energy.
>
> This discovery allows us to understand the dynamics of the construction of the space the universe is made out of: it can bend, stretch, and flow; also, it possesses denser and less denser regions whose importance will later be made apparent for the creation and distribution of galaxies.

Using what can be called the "Ice Cream" analogy for matter and energy, imagine a half gallon tub of ice cream. Next, take a scoop in hand and remove a scoop from the flat surface at the top of the tub. The "hole" that is left on the flat surface of the top can be compared to matter [a proton or electron], while the scoop of ice cream that was taken out can be compared to energy. This relationship is responsible for the famous equation: $E = MC^2$.

A photon is an extremely dense packet of space. Consequently, it displaces the surrounding space, creating a dense region of space around it much like the space surrounding the electron: [not to scale]…

Figure 8.1

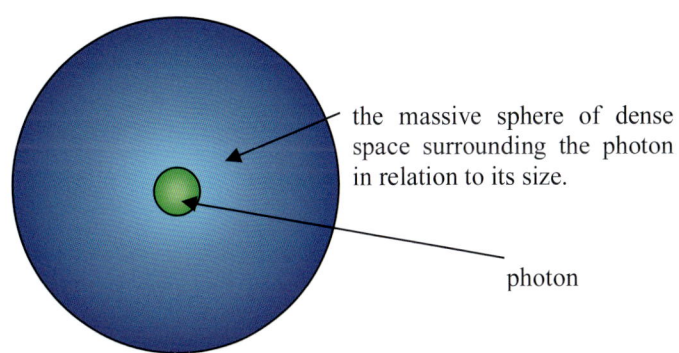

The particle and wave aspects of a photon.

Note, the sphere of dense space that surrounds the photon is massive. Because of space limitations on this page, this drawing is only a representation of the huge sphere of dense space surrounding the photon.

the massive sphere of dense space surrounding the photon in relation to its size.

photon

And again, just like the region of less dense space that surrounds the proton, the region of dense space that surrounds the photon is a function of the three-dimensional space the photon is traveling through. This means that as the photon moves through the three-dimensional space of our universe, the space the photon is passing through bends outwards away from the photon as it approaches, and then back inward as it passes by. Consequently, even though this massive sphere of dense space constantly surrounds the photon and seems to travel with it, *it is the region of space this dense region is passing through that is growing denser as the photon approaches, then less denser as the photon passes by.*

In 1865, [James Clerk Maxwell](#) proposed that light was an electromagnetic wave consisting of an oscillating electric and a magnetic field. That as the electric field expanded, the magnetic field contracted and vice versa. It must be said that this was a brilliant idea, but Maxwell had no idea how these two fields were being generated. But that is all ended. Using the principles of the Vortex Theory, we can now give a visual presentation of how both of these fields are created.

The creation of the photon's *Electromagnetic characteristic* is a result of its spin. Because the electron is spinning when the photon is thrown from the vortex it is spinning too;

The figure below represent two photons seen edge on and moving from left to right across the page. The first photon A; is spinning clockwise; while photon B is spinning counterclockwise. They are both traveling at velocity v, the speed of the vortex from which they were thrown. We recognize this velocity as "C", the speed of light. But it must be remembered that C is in reality merely a function of the velocity of the 3d space in the vortices speeding between protons and electrons.

Figure 8.2

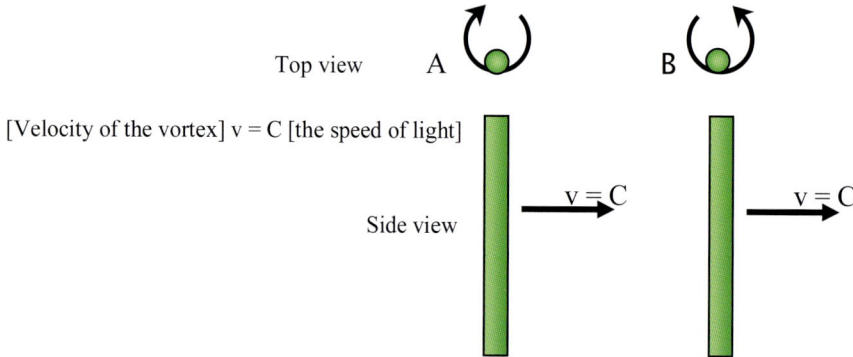

Because both photons are expanding and contracting perpendicular to the direction of travel, it is interesting to note that they can only have one of two spin orientations: clockwise or counter clockwise!

The Perpendicular Expansions and Contractions of the Photon are a most ingenious characteristic of nature.

The photon has to expand and contract outward in long tubular shapes rather than flattened out pancake shapes. If it expanded and contracted in a flattened out "pancake" shape, the velocity of space in front of it would have to exceed the speed of light to get around it creating a "Rip" or "Tear" in the surface of three dimensional space.

Even though the speed of the expansions and contractions of the photon slightly exceed the speed of light, the region of denser space that surrounds it allows this effect. Denser space has a higher elastic modulus that allows it to bend and flex faster.

The reason why the photon even expands and contracts is due to the attempt of its 3d space to blend back into the three-dimensional space from which it came. As the photon races away from the electron it was emitted from, it tries to expand back into the three-dimensional space it originally came from but instead finds that it can only expand in a direction perpendicular to its velocity of travel [see figure below].

The expansion of this dense region of space takes place in a tubular shape allowing the space in front of it to swiftly move around it. As it expands upward and downward simultaneously, the space immediately above and below it is pressed upward and downward respectively. It is the pressure of the surrounding space that keeps the photon from expanding back into the three-dimensional space it is made of. This event occurs when the photon, having expanded as far as its elasticity allows it, finds that the space immediately above and below it is now pushing back down upon it, causing it to contract back upon itself. This contraction continues until the photon is forced back into the condensed spherical shape, beginning its expansion all over again.

As a photon travels through three-dimensional space, its expansion and contraction perpendicular to the direction of its velocity creates its frequency. The length of the beginning and ending of these expansions and contractions creates its wavelength.

Figure 8.3 Photon moving from left to right across page…

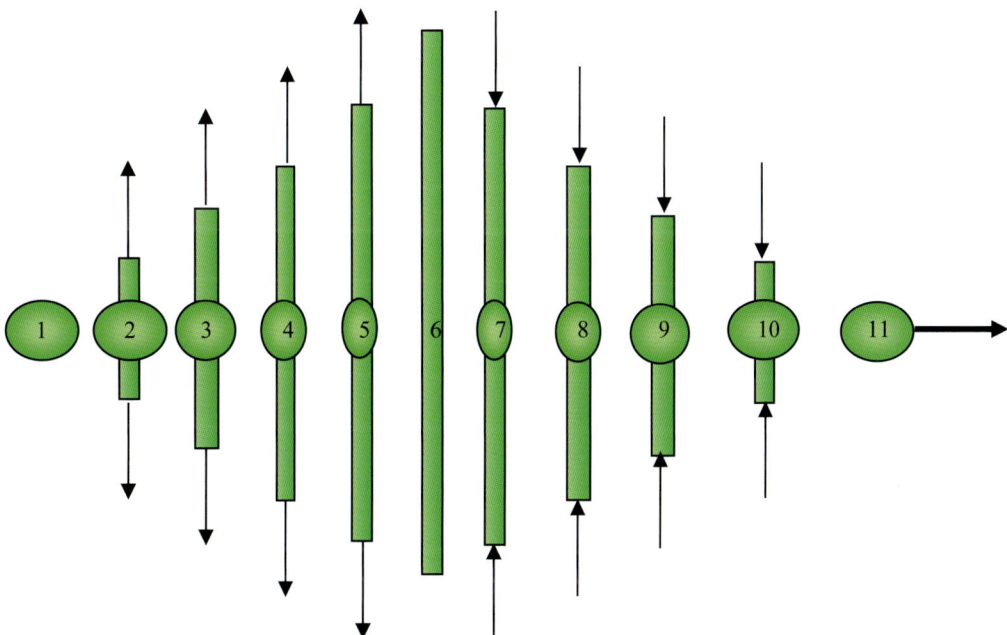

The volume of the denser space within the photon determines its length of expansion and hence its frequency.

The reason for the oscillating exchange of the electric and magnetic fields is a result of the trade off between the changing diameter of the photon. When the photon is in shape #1, it is rotating and its radius is creating maximum rotation of the surrounding space. As the photon expands, it can be seen that this radius grows smaller and smaller, until at #6, the radius is very tiny creating only a very small amount of rotation in surrounding space. However, as it then contracts and goes from #7 to #11, it can be seen that its radius again increases, allowing it to again create a maximum rotation in surrounding space.

But at the same time, it can be seen that as the radius grows smaller, the height of the expansion increases. This expansion creates the electric effect of the photon. At #1, its height is at minimum while the magnetic effect is at maximum. While at #6, the magnetic effect is at minimum while the electric effect is at it its maximum: visually allowing us to now physically observe Maxwell's brilliant deduction.

Note too: it should also be mentioned that the gamma ray possesses so much condensed space that when _seen head on_, it expands and contracts in a six sided "star pattern":

Figure 8.4

But seen side ways, it still looks like this:

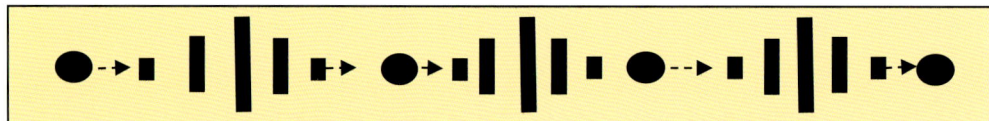

Note: The above was taken from Book 2 chapter 9

See references number [3] on page 79 and number [32] on page 82

Chapter 9
How Subatomic Particles With a Charge of Zero Are Formed

> The difference and great importance between subatomic particles with a charge of ± 1 and those with a charge of 0 [such as the neutron] is presented.
>
> How 0 charged particles such as the neutron are formed allows us to understand later how the space of our universe is in fact turning inside out creating two volumes of space, one in expansion and one in contraction.

In the subatomic world of microscopic space, where the abnormal is normal and the strange is the order of the day, the neutron is the most bizarre inhabitant of all: the neutron is a hole within a hole!

It is easy to see how electrons and protons are three-dimensional holes. This discovery is revealed because their electrical charges are created by three-dimensional space flowing into or out of them. But the neutron has no charge. This means no space is flowing into it or out of it. So how can it be a hole?

The answer is that the neutron is not one solitary hole; instead, it is a hole within a hole! A simply fantastic concept!

The neutron is created when an electron is shoved up against a proton and completely encircles it; or, a proton is hit by an anti-neutrino, its higher dimensional vortex breaks and completely encircles the three-dimensional surface.

Because the electron completely encircles the proton, the space flowing out of the electron is no longer flowing outward into the three-dimensional space of our universe. Instead, its direction is reversed. It is now flowing inwards, directly toward the three-dimensional hole, (the proton) the electron is surrounding. This situation creates an enclosed loop.

The space flows out of the proton and into higher dimensional space; but as soon as it does, it fans outward into a cone shape, is turned inside out, and instantly curls back upon itself creating a tight loop. This tight loop completes the return back into three-dimensional space by flowing directly onto the surface of the encircling electron, forming a fourth-dimensional torus - or donut. A fantastic shape!

Also, just like the proton and the electron, the neutron is rotating around a fourth-dimensional axis of rotation. This axis of rotation is responsible for its ½ spin, and its magnetic moment. The neutron is also surrounded by a sphere of less dense 3d space.

Just like the proton, the neutron also possesses a region of less dense space surrounding it. The proton within originally generated the less dense region that surrounds the neutron.

The region of dense space surrounding an electron does not exist around this electron because the space flowing out of the electron is no longer flowing towards three-dimensional space. Instead, it is flowing in the direction of fourth-dimensional space and is no longer pushing a volume of three-dimensional space back into the three-dimensional universe.

Figure 9.1

Just like all of the previous drawings, the sphere of less dense space surrounding the neutron is massive in comparison to the physical size of the neutron. Hence, this illustration is only a representation.

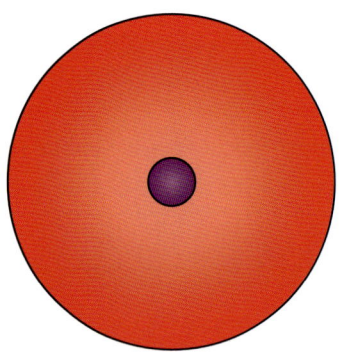

Because this extraordinary creation of nature we call the Neutron is so unique, an attempt to illustrate it was made in the following drawings. Unfortunately, since it is impossible to draw fourth dimensional space, these two dimensional to three dimensional sketches are used:

Figure 9.2

INITIAL CONDITION: anti-neutrino strikes a proton:

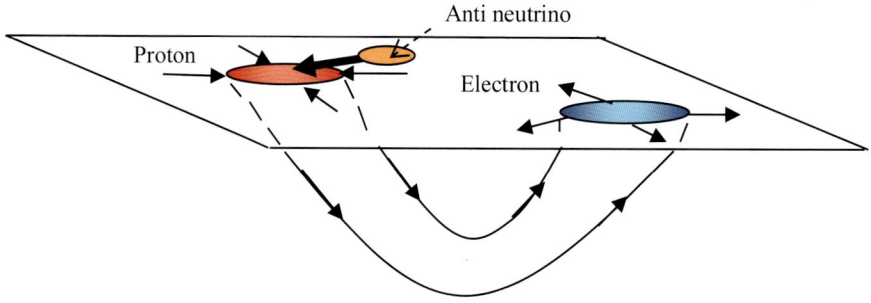

STEP #1: the vortex breaks: [Note, this break is much closer to the surface than seen here.]

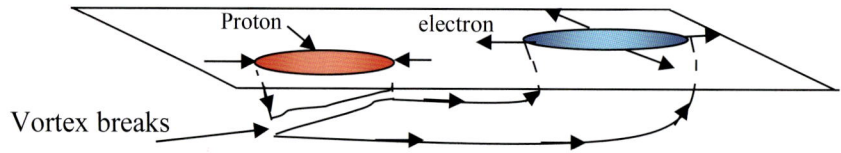

STEP #2: Isolating the proton from the above drawing, and expanding its size, note how the bottom of the vortex begins to curl outward:

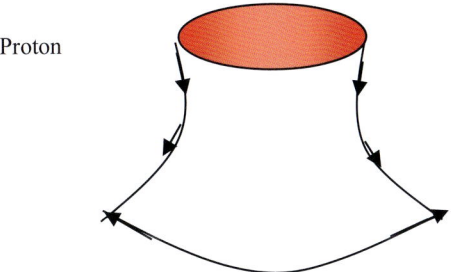

STEP #3: The curl becomes more pronounced as it continues to move upward:

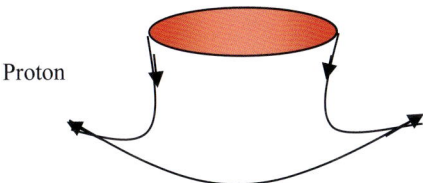

STEP #4: The vortex curls upward at an incredible speed (speed of light) towards the top of the hole we call the proton:

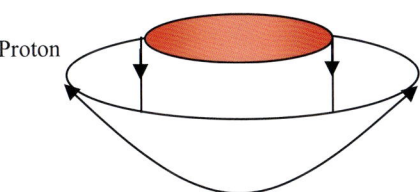

STEP #5: The vortex approaches the top of the hole called the proton:

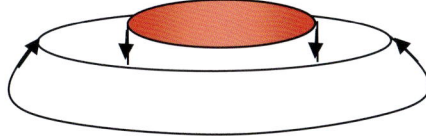

STEP #6: The vortex curls back into the hole called the Proton, forms a torus, the circulating flow begins and becomes a new "particle" science calls the Neutron.

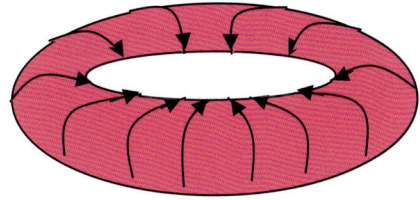

STEP #7: Another new "particle" called a *positron* is created when the end of the vortex attached to the electron reaches the two dimensional surface.

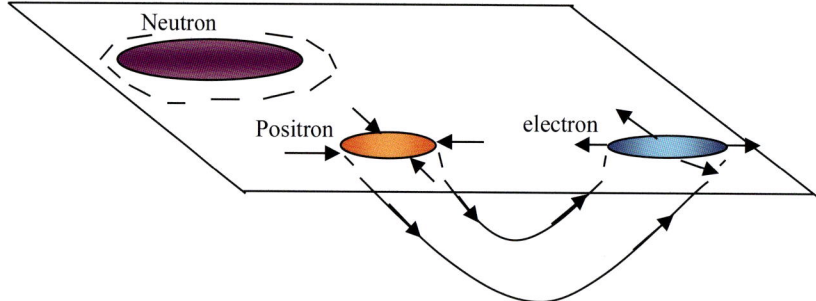

Note how space flows into the hole called the positron; turning it into a "particle" with a charge opposite to that of the electron. The neutron has no charge because none of the space surrounding the neutron flows into it, or out of it.

Also, the higher dimensional vortex is bent into a very tight loop. In this tight loop, the vortex is turned inside out, creating the weak force of nature. Hence, the neutron's "neutral" charge, and the weak force of nature are both explained, clearing up two more of the great mysteries of nature!

But even more important, this illustration also reveals how the one particle the entire physical universe is constructed out of can turn inside out creating a new universe in the process.

The above was taken from Book 2 chapter 7

See reference number [2] on page 79

Chapter 10
How the "FIVE" Forces of Nature Are Created

> The forces of nature are all created by bent and flowing space with the exception of the strong force of nature. This force is created by the exchange of UP and DOWN quarks between protons and neutrons in the nuclei of atoms: they are Yukawa's "virtual particles.
>
> Force of Gravity: its creation reveals at the end of the PROOF how an oscillating universe creates a diverse collection of galaxies instead of a homogenous one.
>
> Electromagnetic force: reveals how space can flow.
>
> Weak force: Reveals how space can turn inside out; and at the end of the PROOF shows how the universe is turning inside out.
>
> Strong force: reveals at the end of the PROOF that there are two sides of space.
>
> Anti-gravity force: shows at the end of the PROOF how matter and ant-matter have separated: with matter creating galaxies, and anti-matter creating areas between galaxies identified as "Dark Energy".

THE FORCE OF GRAVITY

In the theory of Relativity, Albert Einstein believed that matter was made of something and space was made of nothing. And yet, he also believed that gravity was created by bent space surrounding stars and planets. But how can something made of nothing be bent?

Fortunately, the truth is now known. Although space appears to be bent, this is an effect being created by a region of less dense space and is not the cause of gravity; it is not bent space that is creating the force of gravity. The force of gravity is a creation of less dense space.

Most scientists and engineers are told in college courses that the force of gravity is created by a particle called the graviton. Nothing could be further from the truth!

The less dense regions of space that surround the holes ("particles") of matter create the force of Gravity. Einstein's hypothesis that bent space is equivalent to gravity is wrong. *It is less dense space that creates Gravity.* The "bent space" surrounding stars is an *effect* created by the addition of all the spherical shells of less dense space surrounding the protons and neutrons that account for the majority of the mass in stars and other large astronomical bodies.

When we stand upon this planet, this less dense region of space surrounding the Earth distorts the shape of every proton, electron, and neutron within our bodies towards the Earth's center of mass. The collective attempt of these particles to straighten out - pushes us towards the surface of the Earth and becomes the "force" we identify as our weight. Note: *we are not attracted towards the earth*; the particles that atoms are made out of **PUSH** us, accelerate us towards the center of mass of the earth!

Because the red sphere surrounding the Earth represents the three dimensional volume of less dense space that is stretched inwards towards the center of the Earth, the proton is now caught in this less dense region too. But there is now a problem.

As the three-dimensional space surrounding the Earth is stretched towards the Earth, it cannot stretch across the fourth-dimensional void within the center of the proton. Consequently, the side of the proton facing the Earth is stretched towards the Earth while the side facing away from the Earth still retains its spherical shape:

Figure 10.1

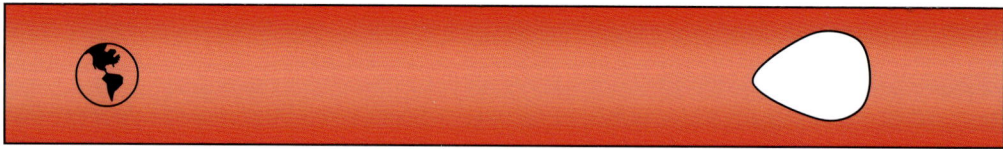

Although the above picture is a poor drawing, you get the idea. You can see how the shape of the proton is distorted in the direction of the Earth. This distortion of the three-dimensional surface of the three-dimensional hole is responsible for creating the force of gravity. It is now understood that gravity is not a "Pull"; instead it can now be seen that surface distortions of the "particles" that make up matter are responsible for the movement of matter towards matter. As can be seen in the diagram below, the distortion of protons in the nucleus of two different atoms are "attracted" to each other by their mutual distortions:

Figure 10.2

(Note: these two holes represent two protons in the nuclei of two different atoms.)

#1

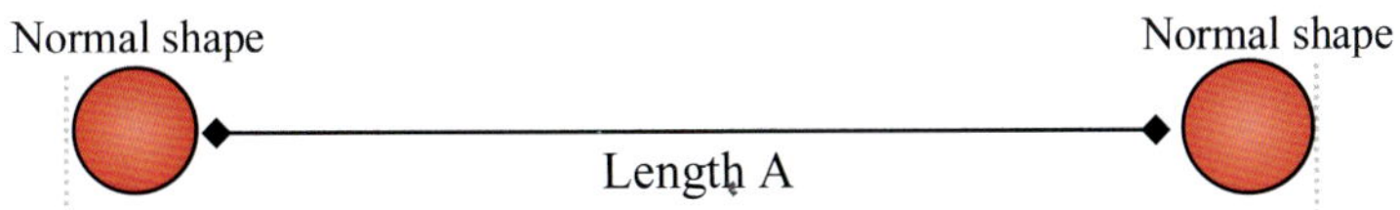

Normal shape   Normal shape

Length A

When two protons are near enough to each other, the less dense regions of space that surround them (these regions are not shown here) overlap, creating an *even less dense region of space directly between them*. This less dense region of space surrounding each sphere stretches the surfaces of the two spheres that are directly opposite each other, but cannot stretch upon the opposite sides of the spheres. This effect distorts the sides of the two spheres that face each other into "pear shapes":

#2

At the very tip or point of the pear shape, space is sharply bent. This stress creates a strain upon the front surface of the hole, forcing the back of the hole to move forward: causing it to move forward.

#3

As the back of the distorted hole moves forward, the spherical shape of the hole is again created: and the proton is now moving at velocity v_1.

#4

However, the instant the hole begins to move forward, regaining its shape, the hole is distorted again; surface stresses at a and b cause such severe distortion that the surface of the hole at point c moves forward again to reduce the strain.

#5

The moving hole returns to the shape of a sphere, and the whole process begins again. But this time, this additional velocity added to the initial velocity now creates an acceleration: $v_1 + v_2 = $ acceleration. And as it continues to accelerate, the velocity gets larger and larger: $\sum = v_1 + v_2 + v_3 + v_4 + n$. Where n = number of distortions.

#6

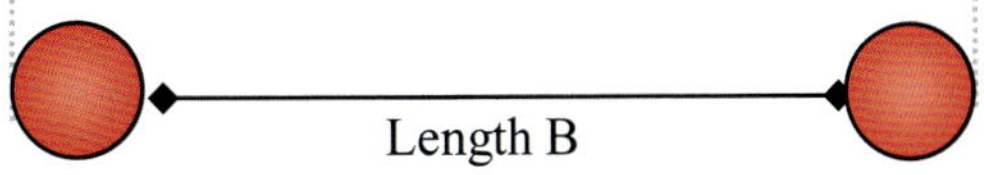

Length B

Length B is now shorter than length A. Equally important, because the distortion of the three dimensional hole distorts the fourth dimensional space within, and subsequently distorts the surfaces of the quarks, the same accelerations are created upon the quarks too. This subject will be discussed in Book 3, *The Quark Theory*.

THE ELECTROMAGNETIC FORCE

The electromagnetic force consists of two parts: the magnetic and electrostatic force. Both of these forces are created by flowing or rotating space: mistakenly called "lines of flux" by people who did not know what they were looking at. The magnetic force is created by the rotation of three-dimensional space; the electrostatic force is created by space flowing into and out of charged particles.

THE ELECTROSTATIC FORCE…

The electrostatic force is created by space flowing into or out of a three dimensional hole. The electrostatic force is created by three-dimensional space flowing into the proton and out of the electron;

Figure 10.3 Electrostatic Force: is created when space flows into and out of the holes we call particles.

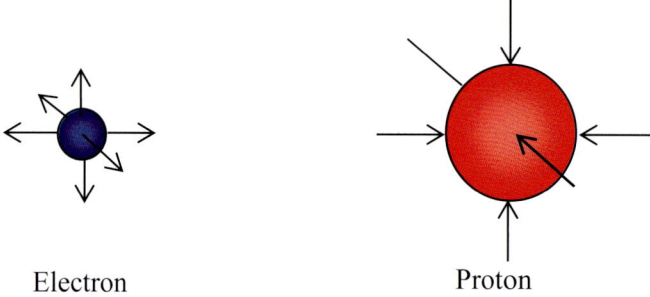

Electron   Proton

THE MAGNETIC FORCE…

Fortunately, for the magnetic force, there is nothing revolutionary to propose. It is fairly easy to understand, that Magnetic fields are created by rotations of three dimensional space around electrons. The spin of electrons creates rotations in space about them creating magnetic fields. It should be noted that the "attraction" and "repulsion" of magnets create the same distortions within protons and electrons. The additions of their spins in bar magnets create the famous lines of flux seen below.

Figure 10.4

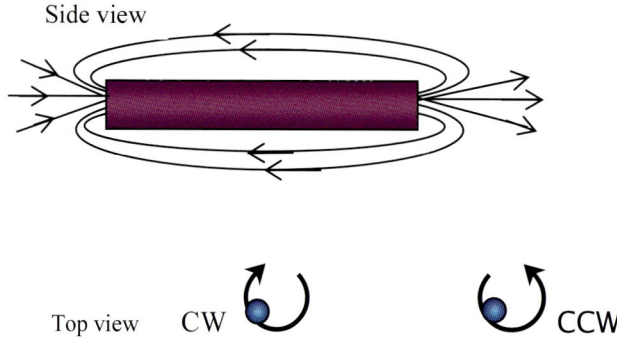

The two spin states of electrons seen above are either clockwise or counterclockwise.

THE ELECTROMAGNETIC FIELD ABOUT A CONDUCTOR…

When current flows in a wire, free electrons in the wire move from one atom to the next. As the electron moves, a point is reached where it breaks contact with one vortex while at the same instant begins to make contact with a vortex from the next atom along its line of travel.

During this brief instant, some of the space flowing outward from the electron is allowed to stretch out to other atoms lying transverse to its line of travel. It is the combination of all of these momentary flows that is responsible for the electromagnetic field about a conductor. The direction of the flows (left hand rule) is determined by the spin of the electrons. The spin of all the electrons is the same because their magnetic moments are aligned to the direction of the electric potential connected to the wire. The strength of the flow – the amount of flowing space – (magnetic field) depends upon the amount of electrons in transit at any particular instant: which is a function of the amount of current in the wire (Coulombs per second).

THE WEAK FORCE

The weak force of nature is not really a force. The weak force of nature is associated with the neutron and how it "decays" into a proton, an electron, and an anti-neutrino. However, the "decay" of the neutron is not caused by a "force", but rather by the break-up of the three dimensional vortex surrounding the proton at the neutron's center! This break-up is caused by harmonics within the tightly bent vortex as it whirls in and out of the proton at the speed of light. These harmonics are created by the space bent inward around the proton opposing the space bent outward around the electron. These two different volumes are diametrically opposed to each other, and seek to escape from each other.

THE TIGHTLY BENT LOOP OF THE VORTEX…

The weak "force", is a result of the higher dimensional vortex being inverted into a tight loop. Because this loop is in the fourth-dimension it is impossible to draw, therefore the following is only a representation.

Figure 10.5

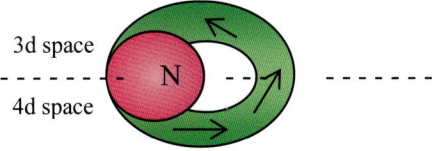

When this tight loop of flowing three-dimensional space is broken, the ends of the vortex are again separated, allowing the proton and the electron to seemingly magically reappear.

What causes the vortex to break is the denser space surrounding the electron trying to move outward and away from the space bent inward surrounding the proton. Because the tight loop is an unnatural bend, the space within the vortex is stretched more on the outside than it is on the inside of the loop. This stretched condition makes it less elastic, decreasing its elasticity.

Within the vortex, this decrease in elasticity makes its inside bend want to flow at a faster rate than its outside bend. This creates a stress between the inside edge of the flow and the outside edge. As the inside of the vortex tries to flow faster, it tries to pull away from the outside of the vortex, and eventually, the construction of the vortex is unable to handle the strain, and it breaks along its outside edge...

Figure 10.6   **Figure 10.7**

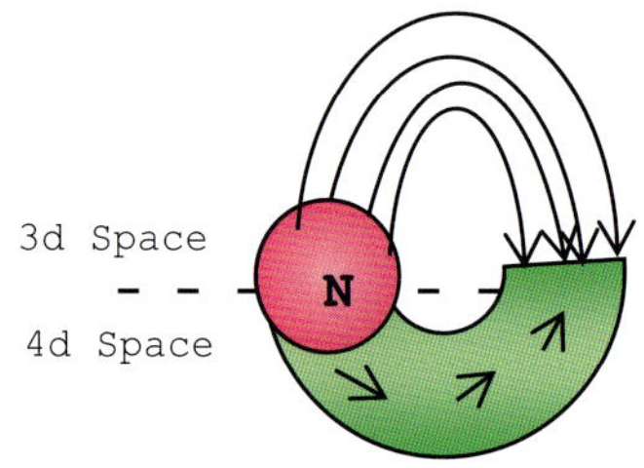

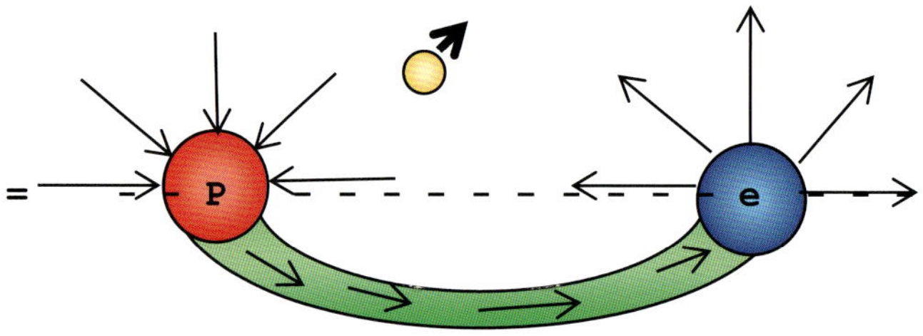

In Figure 10.6, note how the vortex breaks at the neutron; then in Figure 10.7, how the proton reemerges at one end of the vortex while the other end of the vortex becomes the electron; as an invisible electron-antineutrino flies off.

When the vortex breaks, the two ends of the vortex reappear upon the three-dimensional surface as the proton and the electron. The anti-neutrino "particle" is created by the deflation of the volume of the three-dimensional space that the larger neutron filled [explained in Book 3]. The whole process can be compared to twisting a strong spring into a sharp bend, letting go of it, and watching it snap back into its original shape.

Even though the weak force is more of a disturbance than a force, it is interesting to note that both the electromagnetic force and the weak force are related because they are both created out of flowing space.

Also, when the neutron is alone in free space, it only lasts about 10.5 minutes before it "decays" into a proton, an electron, and an anti-neutrino. But when the neutron is within the nucleus of an atom, it lasts much longer. The reason why it survives so much longer is explained in Book 3.

THE STRONG FORCE

PROBLEM WITH THE CURRENT VISION OF THE STRONG FORCE

If the Vortex Theory is correct and quarks are indeed small higher dimensional holes existing within a larger 3d hole, then the current belief that gluons are responsible for the strong force is incorrect. The use of gluons and the principles of Quantum Chromodynamics can only explain how quarks are bound together *within* protons and neutrons [3d holes]; they cannot explain what is holding protons and neutrons together.

This paper proposes that the protons and the neutrons within the nucleus of the atom are being held together by the constant exchange of Mr. Yukawa's virtual particle.

HIDEKI YUKAWA'S VIRTUAL PARTICLE

According to the Vortex Theory, when a neutron first approaches a proton, the less dense space surrounding each "particle" distorts the shape of the other. Then, as both distorted holes attempt to straighten out, they move slightly in the direction of the other. As this process continues, with each additional movement increasing the velocity of the particles, accelerating each in the direction of the other until they collide. As they collide, these two holes try to bend into each other, and as they do, some of the space flowing around and around in the neutron's fourth dimensional [4d] torus tries to flow into the three dimensional vortex of the proton flowing in [4d] space.

This movement of space out of the torus causes it to break. As the torus breaks two ends are created: one end connects to the proton's 4d vortex; the other end emerges back into 3d space, turning it into a proton. At the same instant, the broken end of the original proton's 4d vortex now wraps around and envelopes *it*, changing its vortex into a torus; changing its identity - "metamorphosing" the proton into a neutron; (and the neutron into a proton). However, just as soon as the switch occurs, the process instantly begins all over again, causing the newly formed "particles" to revert to their former identities. This constant switching of identities keeps one "particle" pressed tightly against the other - becoming the strong force of nature.

Figure 10.8

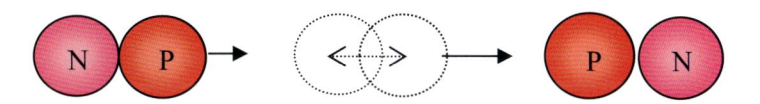

[Note: because the extra three-dimensional space surrounding the larger neutron does not have time to "deflate", but instead is transferred to the proton beside it, no anti-neutrino particle is allowed to form as it does when the neutron is alone in free space. Also, because the extra 3d space surrounding it does not have the opportunity to "deflate", the end of the vortex that is being passed back and forth is much larger than the end that would normally shrink in size to form the electron.]

THE CREATION OF YUKAWA'S VIRTUAL PARTICLE…

Even though the creation of these continual and seemingly instantaneous microscopic metamorphoses is a fascinating and dramatic event, an even more remarkable event is taking place within the proton and neutron. As the proton and the neutron constantly merge together, change identities, and split apart, two quarks are continually being passed back and forth between them.

As the two particles merge together, an up quark from the proton is passed to the neutron, and a down quark from the neutron is passed to the proton. And as this transference occurs, as the two different quarks pass by each other on their continuous journey to the opposite "particle," for a brief instant a virtual pion is created!

In Figure 10.9 below, the proton and neutron are touching; in Figure 10.10, as the proton and neutron begin to change identities, an up quark from the proton and a down quark from the neutron approach each other; in Figure 10.11, as the proton and the neutron merge together, the up and down quarks pass each other creating for a brief instant a **Virtual Pion (with 1/7 the mass of the proton)**; and finally, in Figure 10.12, after the transformation is complete, the neutron and proton have switched identities.

Figure 10.9　　　　　　　　　　　　**Figure 10.10**

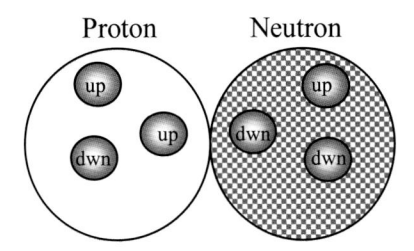

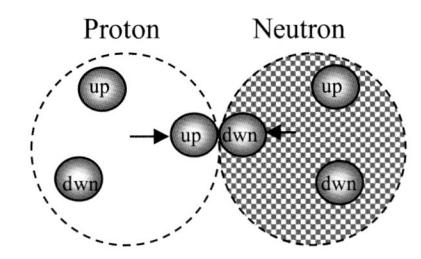

Figure 10.11 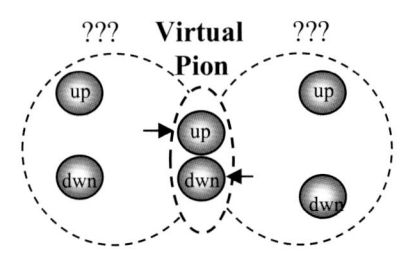　　　　　　　　　　　**Figure 10.12**

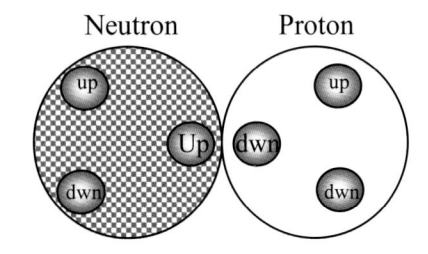

If the Vortex Theory is true, then Hideki Yukawa's proposal of a virtual particle being passed back and forth between the proton and the neutron is correct, and he needs to be posthumously recognized for this great achievement.

THE ANTY-GRAVITY FORCE

Contrary to present belief, there are not just four forces in nature but a fifth one too. Because it is contrary to the other four "forces", this fifth force can be designated the **Anti-gravity Force**. This anti-gravity force is responsible for a number of unexplained phenomena that occur in nature.

Because nothing is presently known about this force, the phenomenon that it is responsible for creating will be revealed. Incredibly, one of the most famous and well known of the effects the anti-gravity force is responsible for is Buoyancy!

We all know about the phenomenon called Buoyancy, this is the reason why boats float in water; however, until now, no one knew why this phenomenon existed.

Buoyancy is important to us all. Without buoyancy, no ship could sail the sea, nor could anyone pan gold. Buoyancy is another one of those phenomenon of nature that is so old, its acceptance is without question. We know that it exists, and because it is a phenomenon that is so common, we don't even think to question it.

But why does buoyancy exist? What is the mechanism in nature responsible for making an object "buoyant"? What makes something float upon the surface of the water? What makes minerals of different densities separate apart from one another in a revolving fluid such as those in a centrifuge?

The answer is something nobody has ever suspected before.

It is not the mass of the object (such as a piece of wood) that causes buoyancy, but the density of the space within the wood! It works like this: the *more* protons and neutrons per cubic inch, the *less dense* the space within; and the *fewer* protons and neutrons per cubic inch, the *denser* the space within. Therefore, it can be said that within a cubic inch of for instance a mineral containing a lesser number of protons and neutrons than another cubic inch of a different mineral containing more protons and neutrons, regions of denser or less dense space are in effect "trapped." This relationship can be seen below in Figure 10.13, A and B: cross sections of rocks.

Figure 10.13

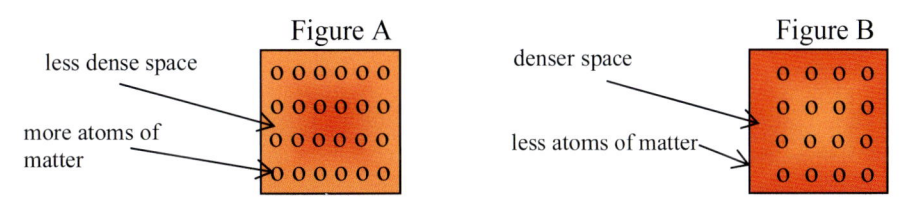

Figure A — less dense space; more atoms of matter

Figure B — denser space; less atoms of matter

In Figure A, more matter per cubic inch creates a region of less dense space in comparison to Figure B. In Figure B, less matter per cubic inch creates a region of more dense space in comparison to Figure A. Within a stream of flowing water, when these different regions of space are being mixed together, or allowed to rotate together, the turbulence of the water allows them to rearrange their locations. The more dense regions of space found in less massive rocks try to move upward, while the less dense regions of space trapped in more massive rocks move downwards. In effect, *when moving*, these denser regions of space possess anti-gravity properties.

It must be emphasized that even though all of the protons and neutrons from both cubic regions of space are accelerated towards the Earth's center of mass equally, when mixed together, it is the denser region of space that is bent upwards and away from the direction of the Earth's center of mass. Consequently, this denser region of space that is trapped within a less dense rock seeks to move upwards and away from the Earth creating a buoyancy effect.

This same effect is produced within the hull of a ship. A region of more dense space is "trapped" within the empty hull of the ship making it push up out of the water.

A hot air balloon rises because *photons are dense regions of space*. When a massive amount of photons are trapped within an enclosure, such as a balloon, the region of air that is constantly exchanging them has in effect entrapped a volume of denser space. Because the surrounding space is less dense, the denser region of space inside of the balloon is accelerated upward.

THE UNIFORM DISPERSAL OF ONE GAS WITHIN ANOTHER...

Another example of anti-gravity effects upon the earth that no-one knows about is the uniform dispersal of one gas within another.

In solids, atoms and molecules are in fixed positions. Between these atoms and molecules are trapped regions of dense or less dense space. It is these dense and less dense regions that create buoyancy. But within gasses, a different situation is occurring.

Within gasses, there are no trapped regions of dense or less dense space. All of the atoms within the gas are pressed against one another. This means the electrons in the outer shells of the atoms are all pressed together but not bonded to each other. Because of this fact, the bent outward regions of space surrounding the electrons in the outer shells of the atoms repel each other.

This repelling effect causes all of the atoms to move apart from one another, to disperse throughout the gas, and seek positions where the effect is at its minimum.

One of the most important questions that 20th Century science has never satisfactorily answered is how come atoms do not merge into other atoms? What keeps the nuclei of atoms from just piling up and creating one monstrous atom?

The answer is the denser space surrounding the electron. This region of denser space repels the denser space surrounding other electrons in other atoms and keeps them apart. In effect, Anti-Gravity keeps atoms from merging into each other. It also is responsible for the acceleration of galaxies away from each other as revealed by the Hubble Telescope

In the late 1990's, a picture taken from the Hubble telescope seemed to indicate that a supernova in a Galaxy located in a distant part of the universe was accelerating away from us at a much faster rate than it should. But how can this be? How can all of the previous spectrographic analysis of the pictures from the last 50 years showing the Red Shift of the Galaxies indicate one result, and this picture indicate another?

The answer to this dilemma can be resolved when it is realized that this picture was taken on the other side of a region in the universe where no galaxies can be seen with earth bound telescopes. This indicates that there is a vast sea of space between the Earth and this distant Galaxy.

In these vast reaches of space where there are few Galaxies, the space is denser. Hence, it tends to cause all of the matter in the Galaxies located on either side of it to be distorted in the direction opposite to it, causing these Galaxies to accelerate away from this vast wasteland towards other less dense regions of space. This phenomenon is responsible for the effect called Dark Energy [created by the presence of anti-hydrogen].

See references number [2] and [3] on page 79, number [25] on page 81, and number 10. on page 85

Chapter 11
The Reason Why Neutrinos Spin CCW and Anti-neutrinos Spin CW Reveals That There Are Two Sides of Space

The explanation for the Conservation of Lepton Number allows us to understand that there are two sides of space. The Conservation of Lepton Number allows us to explain why the neutrino can only spin CCW and anti-neutrinos can only spin CW.

The explanation for the phenomenon of neutrino spin, allows us to realize that that there are two sides of space containing two higher dimensional volumes that our three dimensional space is sandwiched in between.

Because the explanation for the Conservation of Lepton Number is 17 pages long, only a brief synopsis is given here.

SYNOPSIS:

The Conservation of Lepton Number is explained by realizing that there are two sides of space with three dimensional space sandwiched in between. Neutrinos are formed on one side and Anti-neutrinos are formed on the other side.

LEPTON ORIENTATION IN SPACE

A common characteristic that type 1 and type 2 leptons share is their orientation in space. Even though type 1 and type 2 leptons are structured differently, they nevertheless possess distinct orientations in space. For example, neutrinos are all bent into SIDE 1 of space, while anti-neutrinos are all bent into SIDE 2. Electrons, negative muons, and negative taus are all holes constructed on SIDE 2 of space who's *outward flowing space flows into SIDE 1*; while positrons, positive muons, and positive taus are all holes constructed on SIDE 2 whose *inward flowing space flows into SIDE 2*.

In looking at the above orientations, it is important to note that all leptons point in either one of two directions. If positive and negative numbers are assigned to these two directions, and if the direction towards SIDE 1 is called +1 and the direction towards SIDE 2 is called -1, all of the lepton numbers in table can be extrapolated.

Figure 11.1 Lepton orientations in space:

If the direction pointing towards side 1 of space is given a value of +1, and the direction pointing towards side 2 is given a value of -1, the numbers of the twelve different leptons can be determined by the direction of the side they bend or flow into. In the below drawings, notice how the direction of the flowing space determines what the lepton number is:

Figure 11.1

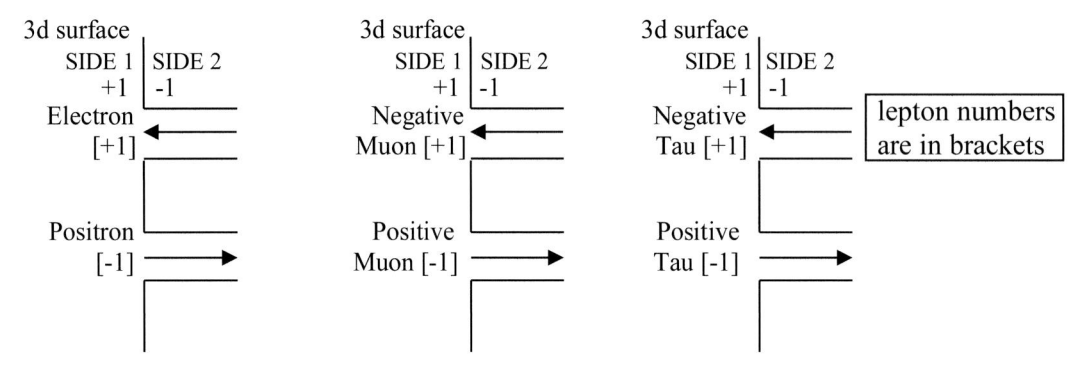

```
    3d surface              3d surface              3d surface
  SIDE 1 | SIDE 2         SIDE 1 | SIDE 2         SIDE 1 | SIDE 2
    +1   |  -1              +1   |  -1              +1   |  -1
  Electron ◄───           Negative ◄───           Negative ◄───        ┌──────────────┐
   [+1]                   Muon [+1]               Tau [+1]             │lepton numbers│
                                                                       │are in brackets│
  Positron                 Positive                Positive            └──────────────┘
   [-1]   ───►             Muon [-1] ───►          Tau [-1] ───►
```

Figure 11.2

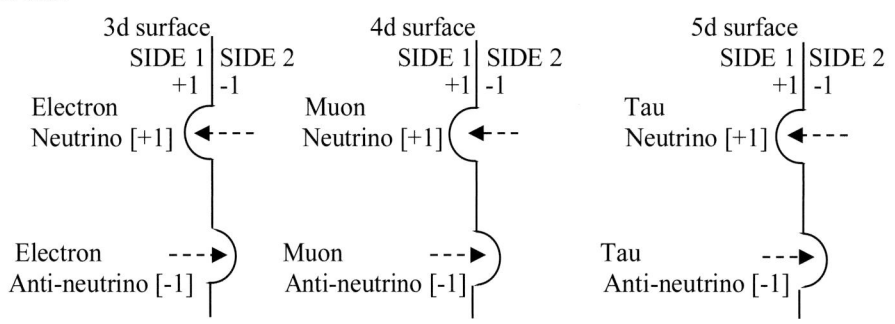

```
    3d surface              4d surface              5d surface
  SIDE 1 | SIDE 2         SIDE 1 | SIDE 2         SIDE 1 | SIDE 2
    +1   |  -1              +1   |  -1              +1   |  -1
  Electron                 Muon                    Tau
  Neutrino [+1] ◄---       Neutrino [+1] ◄---      Neutrino [+1] ◄---

  Electron      ---►       Muon        ---►        Tau         ---►
  Anti-neutrino [-1]       Anti-neutrino [-1]      Anti-neutrino [-1]
```

Observing the above diagrams, it can now be seen that the direction space is bent determines the lepton number. Observe too that the electron, muon, and tau neutrinos, [and their anti-neutrinos] are all created upon the surfaces of different dimensions. These dimensions correspond to the higher dimensional holes the Vortex Theory now proposes to exist within the muon, tau, and their anti-particles.

See reference number [5] on page 79

Chapter 12
There Are Two Volumes of Space, One in Expansion One in Contraction; All Created by One Massive Particle Turning Inside Out

The red shift of the galaxies is being created by the expansion of space. The 1/3 and 2/3 charges of quarks reveal that there are two sides of space one in expansion and one in contraction.

It reveals that there are two volumes of space, one in contraction and one in expansion.

In the 1960s and 70s, the search for and discovery of quarks was the major interest of physics. The discovery of the six types of quarks and their 1/3 & 2/3 charges was considered a major triumph of science. However, nobody has ever explained why quarks possess these particular charges? [What mechanism within quarks generates these charges?]

The scientists of this era merely ASSUMED that these type of charges had to exist to be able to add up to the value of the +1 fundamental charge of nature. For example, inside the proton, because its charge is +1, the three quarks with it have to add up to +1: [2/3 + 2/3 + (− 1/3) = +1]; or within a particle called a pion with a charge of +1: [+1/3 + 2/3 = +1].

However, the fly in the ointment occurs when it is realized that the tiny electron containing no quarks has a charge of − 1, and its anti-particle the positron also has no quarks but nevertheless, still has a charge of +1. So what is going on here?

Well, the problem is not with nature, the problem is with the scientists who are trying to interpret nature. What they did not realize is that a +2/3 charge can also be interpreted to be a charge that is twice the +1/3 charge! Or, simply stated, one charge is twice the other! So how can something like this happen?

It can happen if there are two different volumes of space with one possessing twice the elasticity of the other. In other words, if the density of one volume of space is twice the density of the other, such a situation makes the volume of one vortex twice the volume of the other, making its surface area twice the area of the other, making the charge double the value of the other.

Such a condition in the universe could arise if there are two volumes of higher dimensional space, one in contraction, one in expansion with three dimensional space [the surface of both] trapped in the middle between them.

Figure 12.1

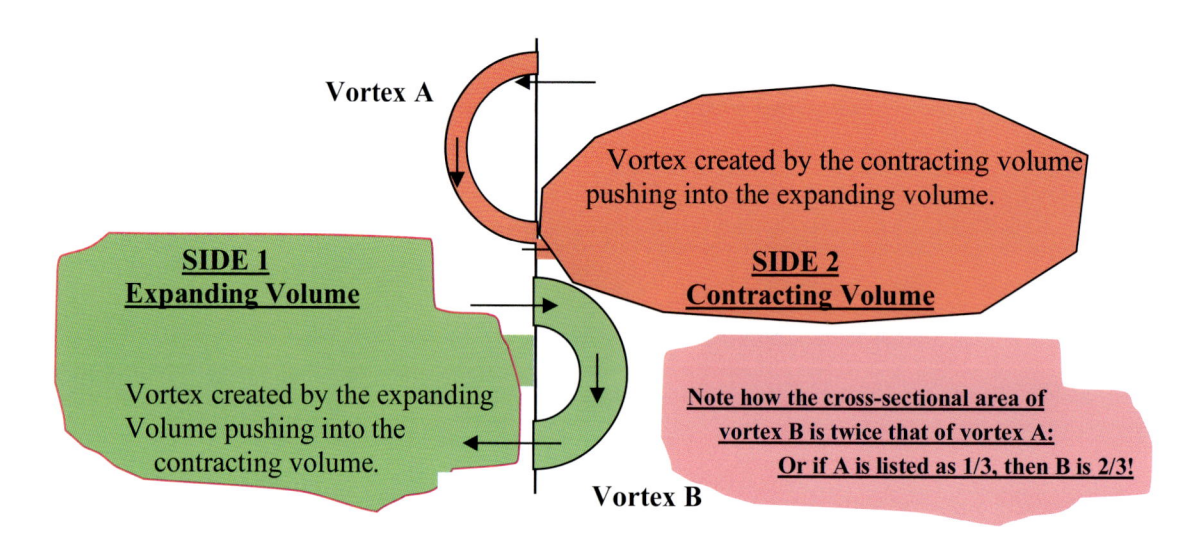

SIDE 1
Expanding Volume

Vortex created by the expanding Volume pushing into the contracting volume.

Vortex A

Vortex created by the contracting volume pushing into the expanding volume.

SIDE 2
Contracting Volume

Note how the cross-sectional area of vortex B is twice that of vortex A: Or if A is listed as 1/3, then B is 2/3!

Vortex B

So how could such an unusual relationship between the different elasticity's of the two volumes of space be created? It can be explained using the analogy of two spheres: one in contraction, one in expansion:

This expansion and contraction can be more easily understood by using the comparison of two spheres – one inside of the other – and sharing the same volume. As the volume of the exterior sphere flows into the interior sphere, the volume of the interior sphere increases and the volume of the exterior sphere decreases. Because what flows out of one volume flows into the other, the rate of expansion is a *one to one ratio*: [as one cubic meter of space is added to the interior of the sphere, one cubic meter is subtracted from the volume of the exterior sphere]. Note in the figures below, that as the interior volume increases, the exterior volume decreases.

Figure 12.2

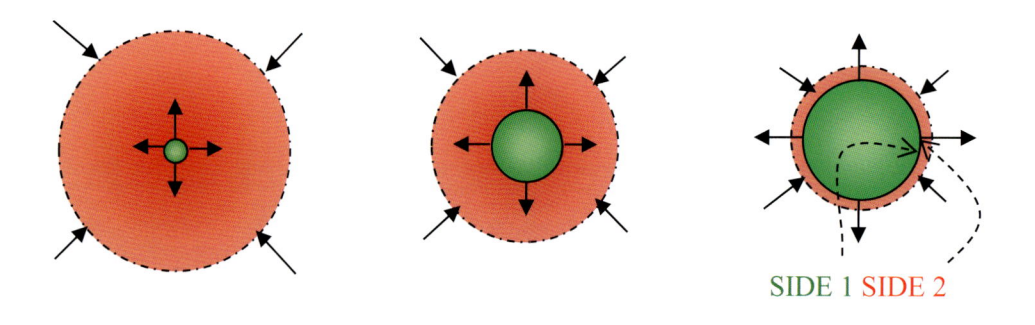

SIDE 1 SIDE 2

Although the flow from the contracting volume into the expanding volume is really a one to one ratio, from the relative perspective of either volume, a different relationship appears. From the relative perspective of either volume, the difference is actually **a *two to one ratio*!** [If the initial interior volume is 10 m^3 and the initial exterior volume is also 10 m^3, as space flows into the interior volume increasing its value to 11 m^3, the exterior volume of space decreases by 1 m^3/sec, decreasing its volume to 9 m^3.] Consequently, because the ratio is two to one, expressed in terms of pressure, it now becomes easier to push outward from the expanding volume [now designated as SIDE 1] into the decreasing volume [now designated as SIDE 2], than it is to push inward from the decreasing volume of SIDE 2 into the expanding volume of SIDE 1. This ability makes it appear as if the elasticity of each volume of space is different.

Transferring this concept to our physical universe and the two volumes of space, this 2 to 1 relationship affects every square meter of the mutually shared surface simultaneously: affecting the elasticity of the two volumes of space everywhere simultaneously.

[Note: this is not speculation. If the difference was not uniform, the flow of the vortices within atoms would not be the same; they would then release photons of differing velocities, (see: The Vortex Theory of Atomic Particles), and we would witness different spectral lines for the same elements in different Galaxies.]

In Figure 12.4 below, the effects created by the two different elasticity's is illustrated: if a vortex of flowing space pushes into the expanding sphere from SIDE 2 creating Vortex A with a flowing volume of 1 m^3/sec; then, Vortex B, Figure 12.3 created by pushing outward from SIDE 1 into the contracting sphere SIDE 2 will have a flowing volume of 2 m^3/sec - twice that of Vortex A.

Figure 12.3   **Figure 12.4**

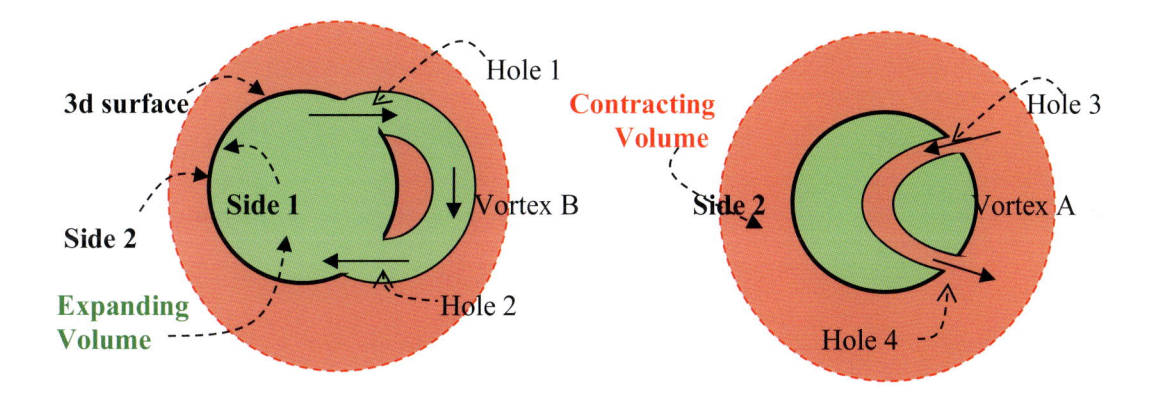

Because space is flowing into Holes #1, and #3, they are considered positively charged; while Holes #2 and #4 that have space flowing out of them are considered negatively charged.

[Although the velocity of space turning inside out is responsible for creating the speed of the flowing space in the vortices [the speed of light], this phenomenon will be discussed later, due to the fact that not only does it appear to be accelerating, but also, due to the incredible possibility that the speed of light only appears fast from our limited perspective!]

Because quarks are fourth dimensional holes trapped within three dimensional holes, and since it will be shown later how the charges on quarks are functions of the space flowing into the three dimensional holes [and not the other way around as is presently believed], then, when two or more quarks are trapped within a three dimensional hole whose three dimensional charge is equal to +1, the flowing volume is divided between them into values that only *appear* to be fractions: [+1/3 +2/3 = +1 (the charge on a positive pion); or, − 1/3 +2/3 +2/3 = +1 (the charge on a proton)].

And even though the charges appear to be divided up into thirds, it is now easy to see the absolute value of the 2/3 charge is really <u>twice</u> the absolute value of the 1/3 charge.

In our search for an explanation for the creation of the universe, it does not seem logical to assume that two different types of space exist with one possessing a greater elasticity than the other. But it does seem logical to theorize that the different dynamics created by the contraction of one volume and the expansion of the other could have created this unique circumstance. The following explanation is one theoretical possibility:

If the three dimensional surface upon which we live was created by a region of higher dimensional space turning inside-out, then the three dimensional space of our physical universe is really a line of demarcation between two volumes of higher dimensional space: one in expansion, the other in contraction.

THE WORK OF DR. STEPHEN SMALE…

In 1959, a strange discovery was made by a brilliant mathematician named Dr. Stephen Smale.

In 1959, Dr. Stephen Smale, a PhD in mathematics and a mathematical genius, published his famous paper entitled: *A Classification of Emersions of the Two Sphere*. Unless one is also a genius in mathematics, neither the title, nor the paper – literally stuffed with abstract mathematical formulas and theorems – seems to make any sense at all. However, to all other mathematical geniuses it presented a simply astounding discovery: that a three dimensional sphere can be turned inside out without tearing its surface!

It was only a few years later, when the mathematics of this paper was used to create visual images, that people could see and appreciate what Dr. Smale had done. These images allowed the drawing of pictures that created a sensation when they were finally published in Scientific American Magazine in 1966.

Years ago when it was first encountered, Dr. Smale's mathematical analysis of a sphere's ability to turn inside out seemed like nothing more than a strange curiosity developed by a brilliant mathematician with a whimsical imagination: something that has no practical purpose whatsoever. However, all that has suddenly changed with the discovery of the *Vortex Theory of Atomic Particles*.

The work of Dr. Smale is suddenly, and absolutely critical in explaining many of the strange phenomenon of the universe: such as the creation of neutral [no electrostatic charges] subatomic particles like the neutron; the creation of the Fine Lines of Hydrogen; and plus, nothing less than the creation of the universe itself! This will be revealed at the end of this book, but first we need it to explain how an electron circling a proton in a hydrogen atom, that is subsequently struck by another electron in another hydrogen atom – ends up turning inside-out! The following little experiment helps…

WHEN LEFT BECOMES RIGHT AND RIGHT BECOMES LEFT!

To better understand what happens when a three dimensional hole turns inside out, try this little experiment. Get an old white T-shirt and a permanent black felt ink pen. Then on the right sleeve draw an arrow that points towards the front of the shirt and mark an "R" beside it; [do it slow enough so that the ink bleeds through to the other side of the shirt], on the left sleeve draw an arrow that points towards the back of the shirt and put an "L" beside it. Now put the shirt on [if you draw the arrows and letters with the shirt on, the ink will bleed through and get on your skin.] When done it will look something like the figure below:

Figure 12.5 [looking from the top down:]

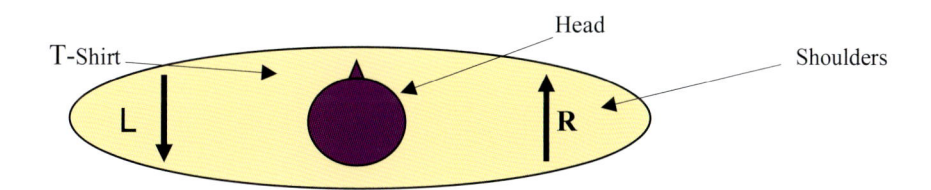

Looking at the above figure, it is easy to see that if one follows the arrow's directions and slowly turns around rotating, one will end up rotating COUNTERCLOCKWISE.

However, when the T-shirt is taken off, turned inside out, and put back on, a different situation arises:

Figure 12.6 [looking from the top down:]

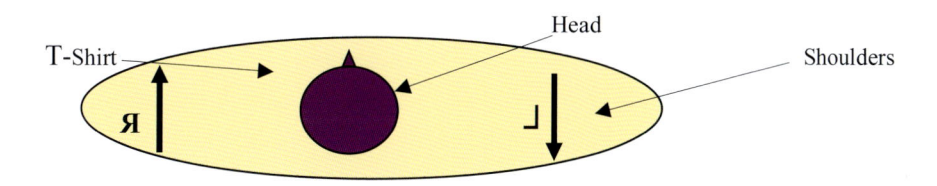

Suddenly everything is reversed. Not only what was on the left is now on the right, and on the right is now on the left, but the arrow on the left now points forward and the arrow on the right now points backwards. If one again follows the directions of the arrows and rotates, the rotation is now CLOCKWISE not counterclockwise! Such is the situation with neutrinos.

See references number 1) on page 85

Also, this work comes from Book 3

Chapter 13
Each Volume Contains at Least 7 Dimensions

> This gigantic "particle" constructing our universe possesses at least seven dimensions. There are possibly more but only later discoveries in particle physics will determine how many.
>
> The layers of leptons and quarks reveal that there are at least 7 dimensions of space in each of the two volumes of space: the one in expansion; and the one in contraction.

If it were not for the PhD Thesis presented at the end of Book 1, *THE VORTEX THEORY OF ATOMIC PARTICLES*, we would never have suspected that "particles" such as protons and electrons are actually holes in space. Also, it would never have been suspected that the quarks existing within protons could also be holes: holes within holes; and, returning to the original hypothesis at the beginning of this book, we would not have suspected that quarks are really a hierarchy of holes and revealing the existence of two volumes of higher dimensional space:

Figure 13.1

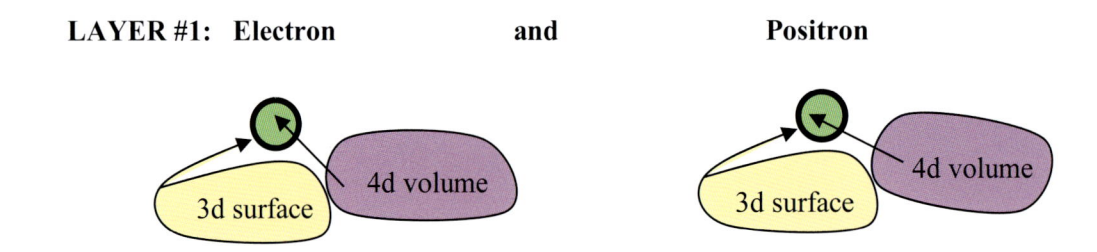

LAYER #1:   Electron   and   Positron

(Diagram: Electron — green circle, 3d surface, 4d volume. Positron — green circle, 3d surface, 4d volume.)

[Note how the electron and positron are 3d holes within 4d space. Both the electron and positron are colored green because they are both formed on side 1 of space: the expanding volume.

Figure 13.2

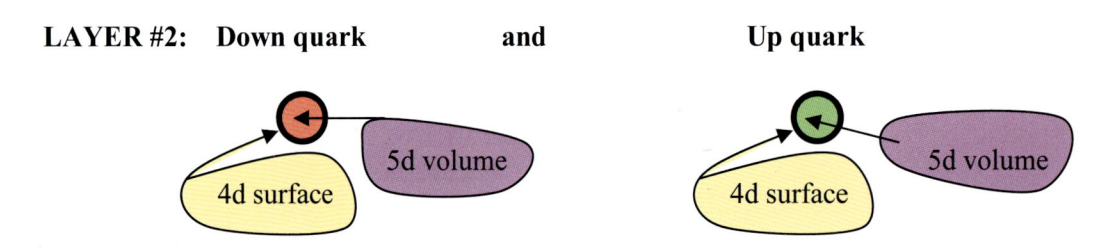

LAYER #2:   Down quark   and   Up quark

(Diagram: Down quark — red circle, 4d surface, 5d volume. Up quark — green circle, 4d surface, 5d volume.)

[Note how the up and down quarks are 4d holes within 5d space. The down quark is red because of its side 2 construction; the up quark is green because of its side 1 construction.

42

Figure 13.3

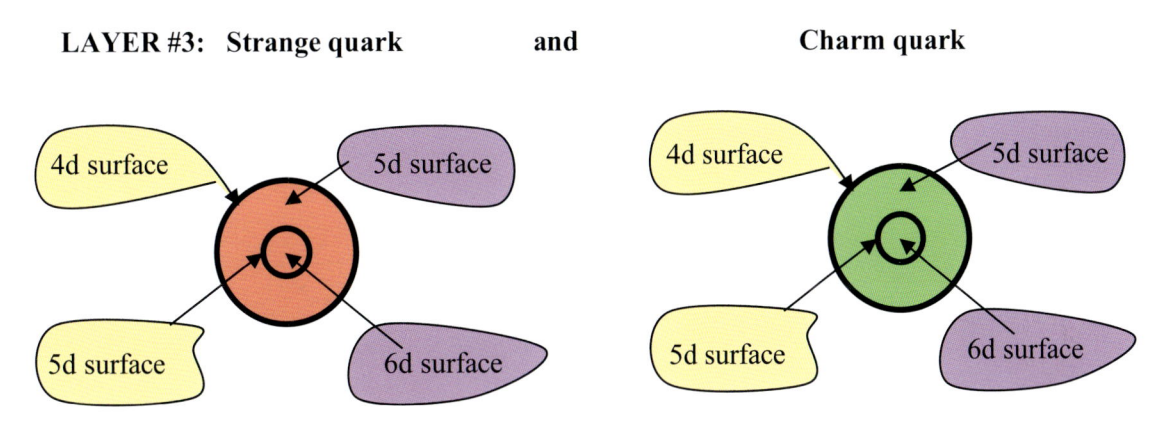

LAYER #3: Strange quark and Charm quark

(Diagram: Strange quark shown as orange circle with 4d surface, 5d surface labels pointing to outer ring, and 5d surface, 6d surface labels pointing to inner circle. Charm quark shown as green circle with the same labeling structure.)

[Note how both the strange and charm quarks are 5d holes within 4d holes. Also, because they are located within down and up quarks they possess the 1/3 and 2/3 charges of these lower dimensional holes they are "sheathed" within. (Strange is side 2; while charm is side 1)]

Figure 13.4

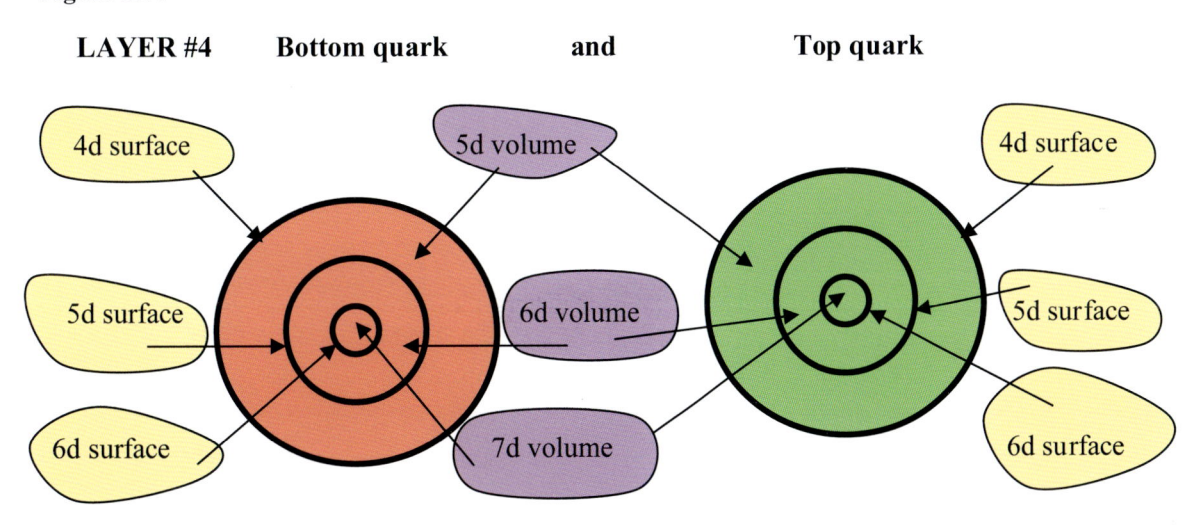

LAYER #4 Bottom quark and Top quark

(Diagram: Bottom quark shown as orange circles with 4d surface, 5d surface, 6d surface labels and 5d volume, 6d volume, 7d volume labels. Top quark shown as green circles with similar labeling.)

Note how both the bottom and top quarks are 6d holes within 5d holes. Also, because they are located within down and up quarks they possess the 1/3 and 2/3 charges of these lower dimensional holes they are "sheathed" within. (Bottom is side 2; while top is side 1)]

From Book 3 Chapter 13 "The Four layers of Matter"

See reference number [38] on page 83

Chapter 14
The Beach Ball Analogy

> To better understand space turning inside out, the "Beach-Ball" analogy is presented:
>
> This analogy reveals, [from a limited certain perspective upon the sphere], how one volume turning inside out appears to be two separate volumes.

Imagine a large thin plastic beach ball twice the size of a basketball; a ball whose outside is colored red and whose inside is colored blue.

Next imagine cutting a slit in its surface, then reaching in and grabbing the inside of the ball and starting to pull the inside out through the slit. As you keep pulling upon it, the inside volume of the ball coming out of the slit with its blue surface becomes larger and larger as the outside of the ball with its red surface becomes smaller and smaller. This continues to happen until the ball has completely turned inside out. The outside of the ball is now blue; and the former red outside is now inside the ball. In a case of juxtaposition, the inside has now become the outside, and the outside has become the inside. In a similar manner, this is what is happening to our universe.

This analogy helps us to understand some of the principles introduced in our next chapter.

Chapter 15
Implications of Space Turning Inside Out; & the Ant and the Basketball!

The implications of space turning inside out are many and shocking to behold. Some of the most important ones are listed below:

a) The giant "particle" is turning inside out. This creates for all practical purposes two volumes of space: one in expansion, one in contraction. [We will come back to this in a moment].

b) The three dimensional surface is in motion. It is expanding outward growing larger and larger.

c) The revelation that one side is in expansion reveals that the red shift of the galaxies is a result of space being in expansion and not galaxies rushing out into the universe on their own.

d) The three dimensional surface we live in is trapped in between the two volumes. It is the mutual surface of the two volumes of higher dimensional space.

e) When looking out into the universe, the perspective we see seems to reveal that we are looking into a gigantic volume. In reality, we are looking outward onto the surface of a gigantic higher dimensional sphere.

f) The curving of light along the surface of the three dimensional surface of the sphere only makes it appear to be a volume like that of a room in the building you are sitting in. Only for short distances does light travel in straight lines. In reality, when we look deep into the universe at objects with a large red shift, their light is coming to us along the curved surface of the universe. [This will be discussed in greater depth in the PROOF revealing a shocking truth about the amount of galaxies that actually exist vs. what we seem to see!!!].

g) The position of the two volumes creates two sides of space. Just as there are two sides of a two dimensional piece of paper, there are two sides of the three dimensional surface.

The ant and the basketball…

Shockingly, and again, as explained in Books 2 & 3, every point in the universe also appears to be the center of the universe! This is a phenomenon created by a surface and not a volume. For example, from the perspective of the volume of space within a basketball, there is only one point that is the center of the volume: the geometric center of the sphere. However, from the perspective of a tiny ant standing upon the outside two dimensional surface of the basketball, as he slowly turns around and looks in all directions as far as he can see, he appears to be standing upon the center of the volume beneath him. And when he shifts his location to any other place on the surface of the ball, he still continues to see the same phenomenon: making it appear as if every point on the surface of the sphere is the center.

This analogy becomes important in our next chapter…

See references number [38] on page 83

Chapter 16
The Three Dimensional Space We Live In Is an Illusion

The three dimension volume of space we live in is an illusion. We live within the three dimensional space that is in actuality, just the surface of the higher dimensional volumes of space the three dimensional surface encapsulates.

We live in a world of illusion. The curvature of light along the spherical shaped three dimensional space we live in makes it seem as if objects in the far reaches of the universe are located upon a straight line away from us. But this is not true.

For many generations too numerous to count, men believed the earth was flat; and so it appears to be from our limited point of view. But we know this is not true. The same for our universe…

From our point of view everything we see in the depths of space with our telescopes seems to be located in straight lines from us. In reality, because we live in a strangely constructed volume of three dimensional space that is really the three dimensional surface of a higher dimensional object, the volume we are looking into is actually circular! Just as the ant stood on the two dimensional surface of the basketball, we stand upon the three dimensional surface of the universe!

It only appears to be a standard three dimensional volume one would encounter in say the room of a building. This discrepancy is caused by the fact that light from distant galaxies travels along the curved shape of three dimensional space, much like a plane flies through the atmosphere of the earth from pole to pole in a curved line instead of a straight one [when all the while, it appears from the pilot's point of view, as if he is flying in a straight line].

See references number 0

Note: we have now garnered enough basic information to explain both the creation and destruction of the universe. And so we begin…

PART IV
THE CREATION AND DESTRUCTION OF THE UNIVERSE!

Chapter 17
FINALLY: The Creation and Destruction of the Universe!

> The creation and destruction of the universe has happened many times in the past. However, it has all happened in a way that we have never encountered before…

With the evidence presented in PART III, we now have all of the information we need to explain how the universe was created and how it will be destroyed.

THE OSCILLATING UNIVERSE

We live in an oscillating universe: one that appears to have been created and destroyed many times in the past. We cannot say when the sequence began or when it will eventually end once and for all, only how this particular cycle began, how it will end, and how it will start all over again. Shockingly, the whole of the universe is one gigantic multi-dimensional "particle" that is turning inside out. This gigantic "particle" possesses at least seven dimensions. There are possibly more but only later discoveries in particle physics will determine how many. This "particle" can be roughly described from our limited perspective as a gigantic multi-dimensional sphere.

DYNAMICS OF THE PARTICLE

The three dimension volume of space we live in is an illusion. We live within the three dimensional space that is the surface of the higher dimension volumes of space the three dimensional surface encapsulates.

The three dimensional surface we live in is trapped in between the two volumes. It is the mutual surface of the two volumes of higher dimensional space.

The explanation for the Conservation of Lepton Number allows us to understand that there are two sides of space.

The position of the three dimensional surface creates two sides of space. Just as there are two sides of a two dimensional piece of paper, there are two sides of the three dimensional surface.

Time does not exist, there is no fourth dimension of "Space-time". There is a fourth dimension but it is a pure dimension of space and contains no time characteristics. Neither does the third dimension of space contain a "Higgs field".

The current belief that space is a void is a mistake. Space is made of something that can bend, stretch, and flow; that can also become denser or less denser.

The giant "particle" is turning inside out. This creates for all practical purposes two volumes of space: one in expansion, one in contraction. [We will come back to this in a moment].

The three dimensional surface is in motion. It is expanding outward growing larger and larger.

The red shift of the galaxies is being created by the expansion of space.

SUBATOMIC VORTICES REVEAL THE EXISTENCE OF 4D SPACE

The sub-atomic particles of matter that all atoms are created out of are really three dimensional holes extending downward into the fourth dimensional volume of the contracting side. The particles of matter and anti-matter are connected by tiny three dimensional vortices of spinning space that extend into the fourth dimensional volume of the contracting side.

EVERYTHING IN THE UNIVERSE IS CREATED OUT OF SPACE

The forces of nature are all created by bent and flowing space; denser or less dense space; with the exception of the strong force of nature. This force is created by the exchange of UP and DOWN quarks between protons and neutrons in the nuclei of atoms: they are Yukawa's "virtual particles. Other particles such as photons of energy are really condensed packets of three dimensional space thrown from the vortices.

Because of the above knowledge, it is now revealed that there are no separate parts of the universe. Instead, all that exists is part of the whole: "All is one!"

THE ILLUSION

When looking out into the universe, the perspective we see seems to reveal that we are looking into a gigantic three dimensional volume. In reality, we are looking outward onto the surface of a gigantic higher dimensional sphere.

The curving of light along the surface of the three dimensional surface of the sphere only makes it appear to be a volume like that of a room in the building you are sitting in. Only for short distances does light travel in straight lines. In reality, when we look deep into the universe at objects with a large red shift, their light is coming to us along the curved surface of the particle. This discrepancy is caused by the fact that light from distant galaxies travels along the curved shape of three dimensional space much like a plane flies through the atmosphere of the earth from pole to pole in a curved line instead of a straight one; when all the while, it appears from the pilot's point of view, as if he is flying in a straight line.

This will be discussed in greater depth in an experiment called, "THE SHOCKING PROOF" for astrophysics later in this book. An experiment revealing an equally shocking truth about the amount of galaxies that actually exist, vs. what think exist!!!

SPACE TURNS "INSIDE OUT": THE BEACHBALL ANALOGY...

Again, from above, "The giant "particle" is turning inside out. This creates for all practical purposes two volumes of space: one in expansion, one in contraction".

To better understand how this is happening, how space is turning inside out, the "Beach-Ball" analogy is once again presented:

Imagine a large thin plastic beach ball twice the size of a basketball; a ball whose outside is red and whose inside is blue. Next imagine cutting a slit in its surface, then reaching in and grabbing the inside of the ball and starting to pull the inside out through the slit. As you keep pulling upon it, the inside volume of the ball coming out of the slit with its blue surface becomes larger and larger as the outside of the ball with its red surface becomes smaller and smaller. This continues to happen until the ball has completely turned inside out. The outside of the ball is now blue; while the former red outside is now inside the ball. In a case of juxtaposition, the inside has now become the outside, and the outside has become the inside. In a similar manner, this is what is happening to our universe.

EXAMPLE OF SPACE TURNING INSIDE OUT: THE NEUTRON

The idea that space can turn inside out is not speculation. It is based upon the creation of the neutron as seen in Ch 9. The difference and great importance between subatomic particles with a charge of ± 1 and those with a charge of 0 reveal those with 0 charge are space turned inside out.

THE DESTRUCTION OF THE UNIVERSE: IN THE BLINK OF AN EYE!!!

Returning our attention to the expansion of the universe, reveals that it will not go on forever. When the volume of space from the contracting side is all fed into the volume of the expanding side, the expansion will stop.

Because all of the three dimensional vortices of the particles of matter extend into the contracting volume of space, when it ceases to exist, all of the vortices and their ends [the three dimensional holes in three dimensional space] will cease to exist. In effect, all matter in the universe will disappear in the blink of an eye. Because there are also higher dimensional vortices between quarks and their anti-particles on the contracting side, these will instantly disappear too!

BUT IT WILL INSTANTLY BEGIN AGAIN!

However, because the expanding space of the expanding volume is accelerating, it cannot come to an instant halt. Although the outside edge has stopped its movement, within the interior at the geometric center of this massive particle, space is still moving towards the outside edge. This movement pulls space apart creating a rip in the shape of a three dimensional hole and the former expanding volume begins to pour into the hole: and the universe begins again.

The former expanding side now becomes the contracting side and the volume of space moving out of the center becomes the expanding side.

The beginning of the outside twist in space begins as a tiny pinpoint and expands outward at the speed of light forming matter with it [as will be shortly explained]. This beginning is what has mistakenly been referred to as the "Big Bang".

THE "RE-CREATION" OF MATTER

As the inside volume of space begins to expand outwards becoming larger and larger, but since it is connected to the space it is expanding out of, resistance to its expansion is now encountered. This resistance is created when the expanding volume becomes less dense due to its rapid expansion. Because the space of the expanding volume and the contracting volume are connected to each other, the space between them is stretched. This creates resistance. This resistance tries to pull the expanding volume backwards. This backwards pull cannot pull the expanding volume back into the contacting volume. However, its resistance creates distortions in the expanding volume's surface. This resistance creates an enormous amount of three dimensional holes upon its three dimensional surface; creating in turn, a new generation of vortices that extend downward into the higher dimensional space beneath the surface and then begin to return. As these vortices return to the expanding three dimensional space of the surface, the whole process of matter anti-matter creation begins again, and a new version of the universe is born again.

The first particles created appear to be the lambdas. The Lambda possesses 1.1 times the proton's mass, has a 63.9% chance of decaying into a Proton and a charged Pion [a particle containing two quarks]. This is a good possibility because the charged Pion then decays into a muon and a neutrino, that in turn decays into an electron and two neutrinos. And since protons [and anti-protons], electrons [and positrons], and neutrinos [and anti-neutrinos] dominate the particles of the universe, the Lambda is a good possibility that massive amounts of these were the first particles produced by the backward resistance created by the contracting side upon the expanding side. Their decay products then created the particles that today populate the universe. However, at the time of this writing this is supposition.

CREATION OF THE CMB BACKGROUND RADIATION

In the very early creation of particles of matter, the close proximity of matter and anti-matter created annihilations, creating massive amounts of photons that eventually became the background radiation of the universe.

If the three dimensional space we live in was nothing more than a volume, these photons should have gone out into the universe never to be seen again. However, because the three dimensional space we live in is the outside surface of a gigantic higher dimensional spherical particle, three dimensional space is its surface; and just as the atmosphere of the earth is a volume that is curved around the spherical surface of the earth, our three dimensional space is also curved around the volume of this higher dimensional particle.

Consequently, the CMB radiation that was created in the early universe, has come around the backside of the surface of the higher dimensional sphere, and is now coming back at us!

This phenomenon also possesses a profound revelation for the light from distant galaxies that is also coming to us from the "backside" of the universe and will be expounded upon in the PROOF FOR ASTROPHYSICISTS later in this book.

CREATION OF GALAXIES AND REGIONS OF DARK ENERGY

When the three dimensional space of the surface was pulled into the three dimensional particles, spherical regions of less dense space surrounding them was created, creating nuclear gravity. At the other end of the vortices that three dimensional space was flowing out of, the surrounding space was pushed outward creating denser regions of space that surrounded them, creating anti-gravity. And then later, when the separation of the particles became greater, and they were no longer close enough to annihilate each other, the anti-gravity particles began to move away from those particles generating nuclear gravity. This created two types of clustered particles, those with gravity that eventually formed galaxies; and those of anti-gravity that moved to the empty regions of space between the galaxies and became what science now mistakenly calls "Dark Energy".

ABNORMALITIES IN THE INCOMING VOLUME OF SPACE FROM THE FORMER CONTRACTING SIDE CREATES REGIONS WHERE GALAXIES FORM

When the space from the former expanding side now begins to pour into the volume of the newly formed expanding side, it possesses discrepancies in its surface. Where galaxies once existed, the less dense space that marked their mutual gravity is still there. In effect, these less dense regions create depressions in the surface, making it easier for new holes in space to form; making areas for new galaxies to form in these regions.

These depressions in space are responsible for creating diversity in the new universe. Where a homogenous distribution of galaxies should have formed, a heterogeneous distribution of galaxies are now seen [such as: "Strings" of galaxies; The "Wall" of galaxies; and massive "Super-clusters" of galaxies; etc.].

These depressions are also responsible for creating abnormalities in the Cosmic Background radiation of the universe as reported recently but misunderstood by Sir Roger Penrose of Oxford University in England.

Does the creation of matter continue? Or rather when does it stop? The answer to this question is unknown at the time of this writing. However, it is theorized that the volume of vortices extending into the contracting side lessens the resistance to the expansion and a point of equilibrium is eventually reached where no new matter is formed.

THE END OF THIS CYCLE!

Again, when all of the space from the contracting volume has flowed into the expanding volume, all of the vortices will disappear in the blink of an eye along with all matter, ending this cycle. Will it begin again? Again, the answer is unknown at the time of this writing.

PART V
SOME INTERESTING ILLUSTRATIONS?

Chapter 18
Analogous Drawings Used to Illustrate the Beginning of the Universe…

> Originally the following drawings were meant to be visual aids for the creation and destruction of the universe; but were later edited out because they are for the most part merely analogies that might confuse. However, later I was advised that they might still help those who were struggling with ideas such as space turning inside out. So here they are…

A SCHEMATIC DRAWING: SPACE TURNS INSIDE OUT

Although it is impossible to draw higher dimensional space, the following analogy is sufficient to explain the scenario for what can only be called "Space Turns Inside Out". Also, it must be understood that the Rip or Tear [the hole] in its surface depicted below and the size of the universe depicted here are greatly out of proportion.

Figure 18.1 Note again, this is only a two dimensional schematic drawing: Volume 2 [green] is expanding; while Volume 1 yellow is contracting.

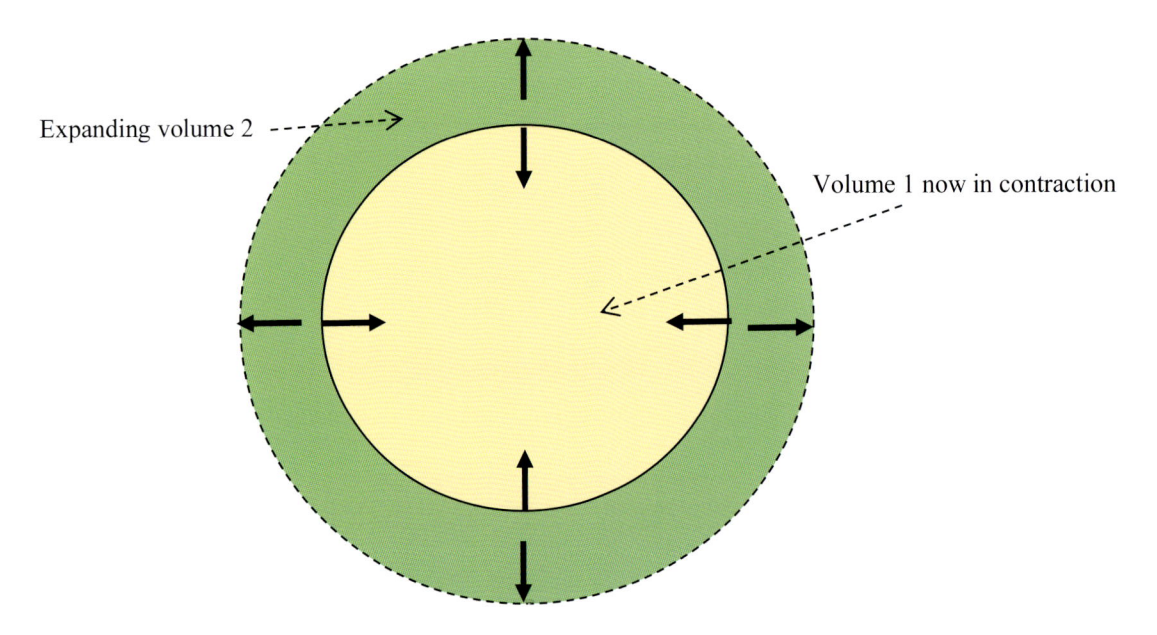

Although it is impossible to see because it is impossible to draw, each volume contains seven dimensions. The three dimensional space we live in is the line where the green meets the yellow. Our three dimensional space is really the mutual three dimensional surface of the fourth dimensional volumes of both "particles".

Figure 18.2 As the space from the contracting side flows into the expanding side, the contracting volume grows smaller, and the expanding side grows larger…

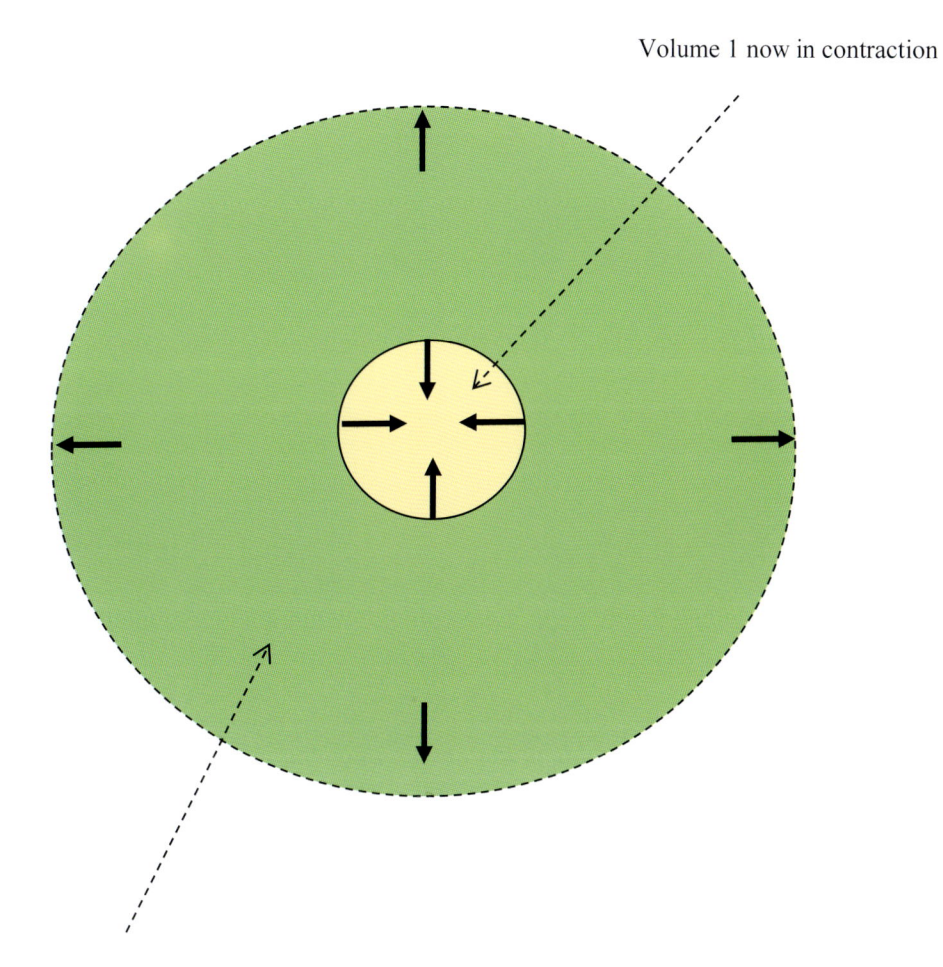

Volume 1 now in contraction

Expanding volume 2

Figure 18.3 When all the space from the contracting side has flowed into the expanding side, the outside edge of the expanding side stops moving. However, because of the length between the edge of the outside of the circle and its center, the center of the sphere continues to try to move outward towards the edge. This outward motion pulls upon the spot directly in the geometric center. When this happens, space is suddenly pulled apart, moving outside of the former expanding volume and begins a new rapid expansion as is seen below:

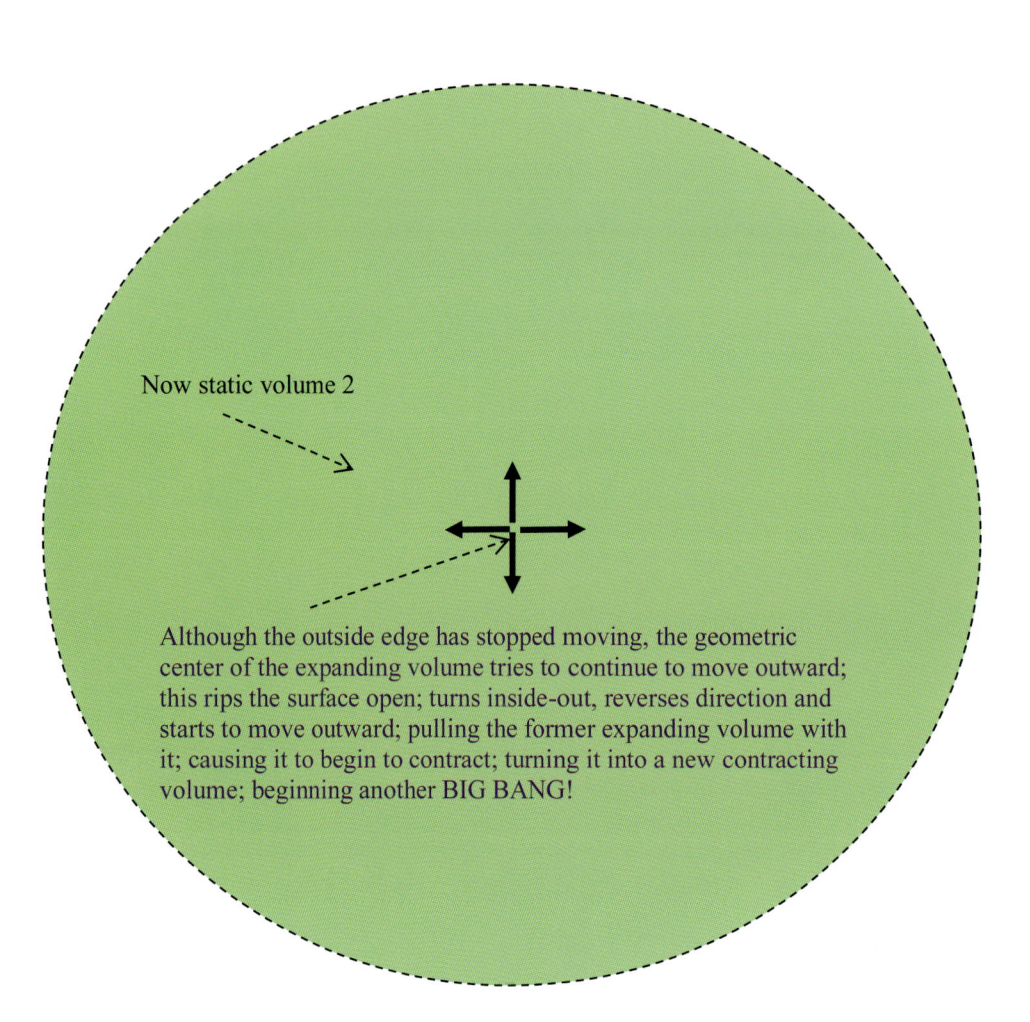

Now static volume 2

Although the outside edge has stopped moving, the geometric center of the expanding volume tries to continue to move outward; this rips the surface open; turns inside-out, reverses direction and starts to move outward; pulling the former expanding volume with it; causing it to begin to contract; turning it into a new contracting volume; beginning another BIG BANG!

Figure 18.4 The outward motion of the space immediately surrounding the expanding volume's geometric center created a spherical rip in space; causing it to suddenly twist inside out of its volume and began to rapidly expand outward. The situation is now reversed from Figure 18.1. Volume 2 that was in expansion begins to contract, and the new Volume 1 begins to rapidly expand outward dragging Volume 2 into the breach. The whole of Volume 2 now reverses direction and moves inward and the universe begins its regeneration.

Also, as will be shown later, because space turned inside out, Volume 1 begins to rotate in the opposite direction to Volume 2. According to the explanation of the Asymmetric Parody of Neutrinos in Book 3, the Volume in contraction now rotates counter-clockwise rotation while the Volume in expansion rotates clockwise.

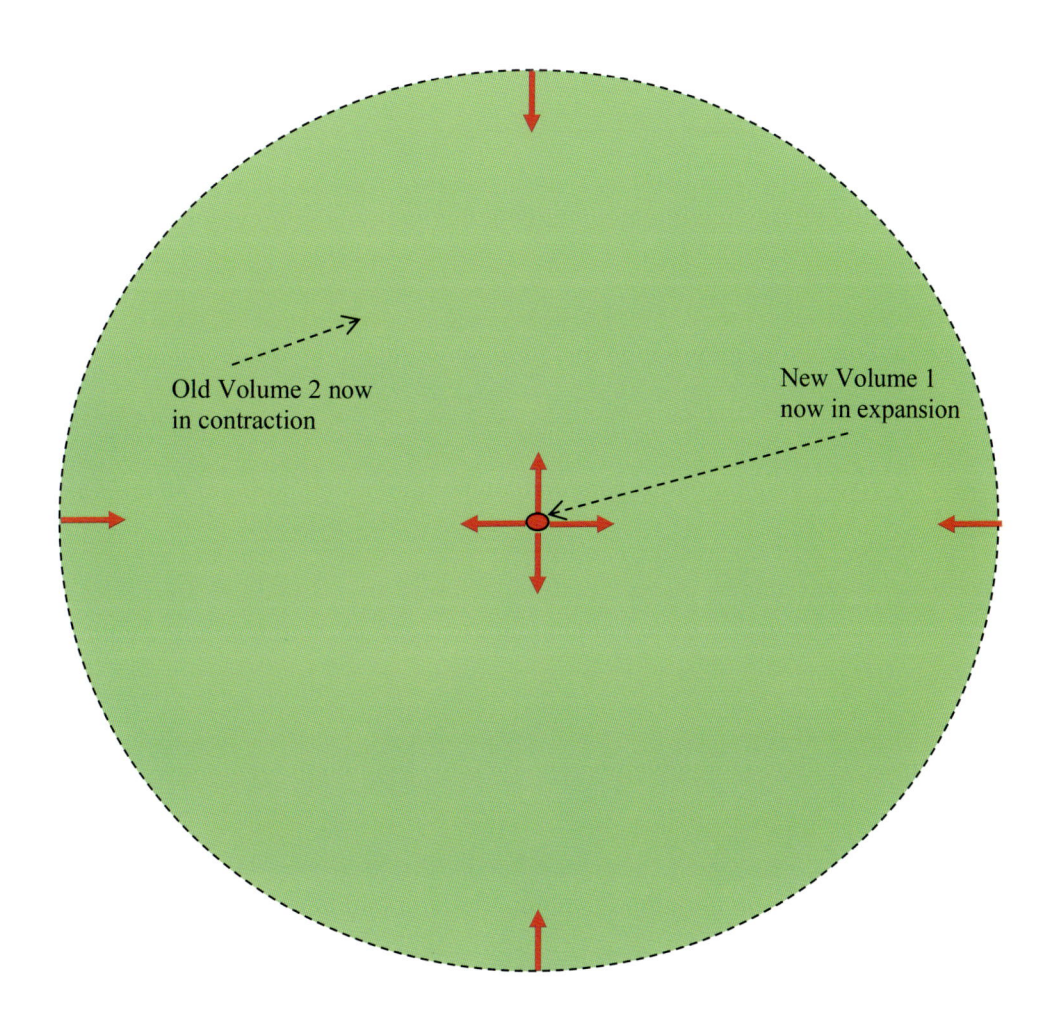

THE RATE OF THE EXPANSION

The key to the rate of expansion of side 2 into side 1 is found in the speed of the space flowing in the vortices. It is no accident that they are flowing at the speed of light: C. Because three dimensional space is "sandwiched" between the higher dimensional space of the expanding side and the higher dimensional space of the contracting side, the difference in their expansion rates is found in the speed of the vortices: C, the speed of light.

> NOTE: In the 2005 thesis for which a PhD was granted by the Russian Ministry of Education, not only was it revealed that matter - protons and electrons - are three dimensional holes in the surface of fourth dimensional space; they are connected by tiny whirling vortices that extend into fourth dimensional space; that the space of the universe is made of something and behaves - from our point of view - as a single particle of absolutely incredible, size!

Chapter 19
Analogous Drawings Used to Illustrate How the Particles of Matter Are Created

> Again, all the following drawings are only <u>schematic drawings</u> made to illustrate how the particles of matter are created.

Because the vortices flow at the speed of light, creating photons that move at the velocity of the vortices; and these two phenomenon reveal that the rip in space expanded outward at the speed of light.

As volume 2 [the volume in expansion] expanded outward at this incredible velocity, volume 1 [the volume now in contraction] tried to resist the expansion. This resistance resulted in the instant creation of billions of trillions of trillions of depressions in the three dimensional surface in expansion. These surface depressions deformed downward and inward into the higher dimensional space of its volume; creating vortices into which three dimensional space flowed into and out of. The end which three dimensional space was flowing into became the positive charge of the "particle"; while the end where three dimensional space flowed out and back onto the three dimensional surface became the negative charge of this "anti-particle". [Note: because we cannot look into fourth dimensional space to see the vortex, and since these three dimensional holes are so minute, these three dimensional holes at the ends of the vortices appear from our point of view to be individual particles.]

Schematic of the cross section of the expanding volume of space, as seen from higher dimensional space. Note how the ends of the vortices create particle anti-particle pairs. [The arrows represent space flowing into the positively charged particle and out of the negatively charged particle.]

Figure 19.1

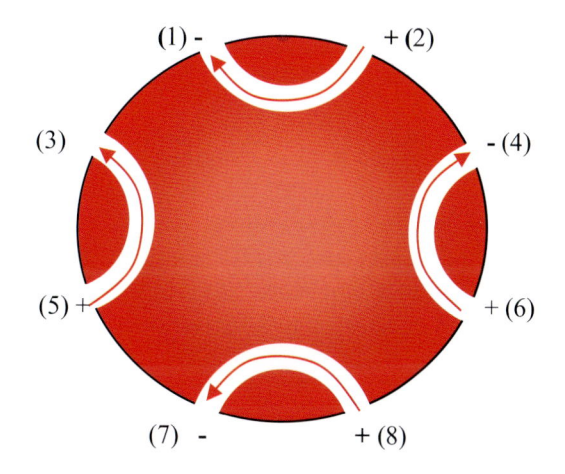

Although the above drawing represents only four examples of particle anti-particle pairs, it must be understood that the expanding volume of space would literally be pock-marked like an enormous golf ball with billions of trillions of trillions of holes in its surface as seen in below:

[Schematic of the cross section of the expanding volume of space, <u>as seen from the perspective of three dimensional space</u>. Note how only the ends of the vortices that create particle anti-particle pairs can now be seen. The positive and negative signs now represent the electrostatic charges, the "flowing space" moving into and out of the particles as perceived from the perspective of three dimensional space:

Figure 19.2

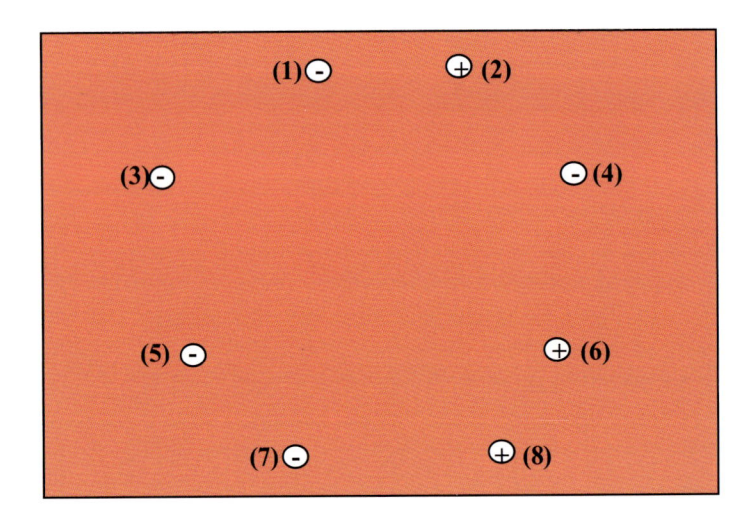

THE EARLIEST PARTICLES

What were the earliest particles created in this "Big Bang"? This is a question that contemporary science cannot answer. It is a mystery for which there is no "epoch".

Also, as explained previously, contemporary science's proposal that a massive volume of quarks and gluons eventually came into being is also wrong. Again, as mentioned previously in the Vortex Theory Books 2 & 3, it was discovered that quarks are higher dimensional holes [fourth, fifth, and sixth, dimensional holes] trapped within the fourth, fifth, and sixth dimensional volumes of space all existing within three dimensional holes and cannot "escape" to the three dimensional space on the surface.

In an effort to explain the creation of the protons and neutrons, contemporary science postulated that a gigantic cloud of quarks and gluons existed that later came together and became the nuclei of hydrogen atoms. However, as just mentioned, quarks cannot exist separately on their own, so this proposal of contemporary science just isn't true. Consequently, a completely different scenario must exist to explain the creation of the protons, neutrons, and electrons that made up early hydrogen and its isotopes; that then condensed to form the stars and galaxies.

Although this question will be better answered by a next generation of physicists trained in the science of the cosmology of the Vortex Theory, it can be theorized here that these early particles were created when space turned inside out creating two volumes: one in expansion, one in contraction. Previously, it was proposed how the sudden increase in the volume in expansion encountered drag, or resistance from the volume previously in expansion. This resistance sought to pull this new expanding volume back into the other volume. However, instead of pulling it all back in, this resistance created indentations or depressions in its three dimensional surface that morphed into vortices where space flowed into one end and exited the other.

The depth and size of these depressions depended upon the strength of the force side 2 [the volume now in contraction] had upon side 1, [the volume in expansion]. These depressions were ultimately responsible for creating the matter and energy of the universe. The depressions or holes created the matter of the early universe, and the volume of space that came out of the holes created the energy of the early universe [the CMB background radiation].

In an effort to determine just what these early particles were, we look at what they created: protons, electrons, photons, and neutrinos. In studying the particles that could have decayed and created all of the above, we look at the more massive baryon particles and a most promising candidate, the Lambda.

THE LAMBDAS?

Discussed earlier, the lambda possesses 1.1 times the proton's mass, has a 63.9% chance of decaying into a Proton and a charged Pion [a particle containing two quarks]. This is a good possibility because the charged Pion then decays into a muon and a neutrino, that in turn decays into an electron and two neutrinos. And since protons and electrons dominate the particles of the universe, the Lambda is a good possibility that massive amounts of these were the first particles. Because some of the neutrinos will hit protons creating neutrons, it could also account for the eventual creation of Deuterium and Tritium. However, since the Lambda also has an additional 35.8% chance of decaying into a neutron and a neutral Pion, this creates a higher percentage of Deuterium and Tritium than appears to exist in the universe. However too, this observation is based upon the possibility that the neutrons would be created in close proximity to protons. Causing them to be almost instantaneously paired with protons and would not exist singularly by themselves. Not allowing them to eventually decay normally into a proton, electron, and anti-neutrino as usually happens in approximately 10.5 minutes.

It must also be understood that the Lambda will also be created with its anti-particle. This anti-particle will in turn decay exactly like the regular particles except they will create the anti-particles of the previous decay particles. The decays of these anti-particles of these heavy baryons will also happen very rapidly in about 10^{-10} or less; resulting in the ultimate creation of anti-protons and positrons.

And finally, because all of these particles will be created in close proximity to each other, some of the particles and their anti-particles will annihilate creating photons. This will ultimately create a massive "cloud" made out of protons, anti-protons, electrons, positrons, photons, and neutrinos. The photons will eventually be called the cosmic background radiation [the CMB] and the neutrino background: the CVB neutrinos.

A HIERARCHY OF PARTICLES?

Another possibility that must be discussed is the creation of a hierarchy of particle creation. It is a possibility that the initial difference in the densities of the space between the side in expansion and the side in contraction was great, causing the following hierarchy of particle creation: #1 Pentaquarks; then #2 Baryons; next #3 Mesons; and finally #4 leptons [specifically electrons and positrons]. The creation continued until the large amount of vortices flowing into the contracting volume lessoned its density and neutralized the difference in densities between it and the expanding volume, making them equal. The Pentaquarks, Baryons, and Mesons then decayed.

The argument *against* this type of creation scenario comes from realization that if it indeed happened, the CMB radiation would be diverse. The argument *for* this scenario is the great amount of free electrons that are theorized to exist in the universe: one per one cubic meter of space. However this is an argument that will have to be left for future cosmologists.

Chapter 20
What Ended the Initial Creation?

What caused the initial creation of matter to end? Creation of hydrogen and anti-hydrogen; the problem with the cosmic microwave background radiation: the CMB radiation.

The initial creation of the sub-atomic particles the matter the universe is made out of would have halted when the volume of space in contraction would have no longer created resistance to the volume of space in expansion. So what caused this?

It is theorized that the end of the initial creation would have happened when an adequate amount of the vortices flowing into and out of the volume in contraction displaced enough of this volume to make it less dense, lowering its resistance to the volume in expansion.

For example: In Figure 20.1 #A below, the line of arrows represents the contracting volume trying to halt the increase of the expanding volume. In #B, shows the effect of one such arrow [dotted line] pulling back upon the expanding volume. As it continues to pull back, in #C, a vortex of flowing space begins to form. In #D, the vortex of flowing space has formed. The inward flow creates the positively charged particle, while the outward flow creates its anti-particle. The extension of the vortex into the contracting volume changes its density, making it less dense: lessening its "pull" upon the expanding volume.

Figure 20.1

#A

#B

#C

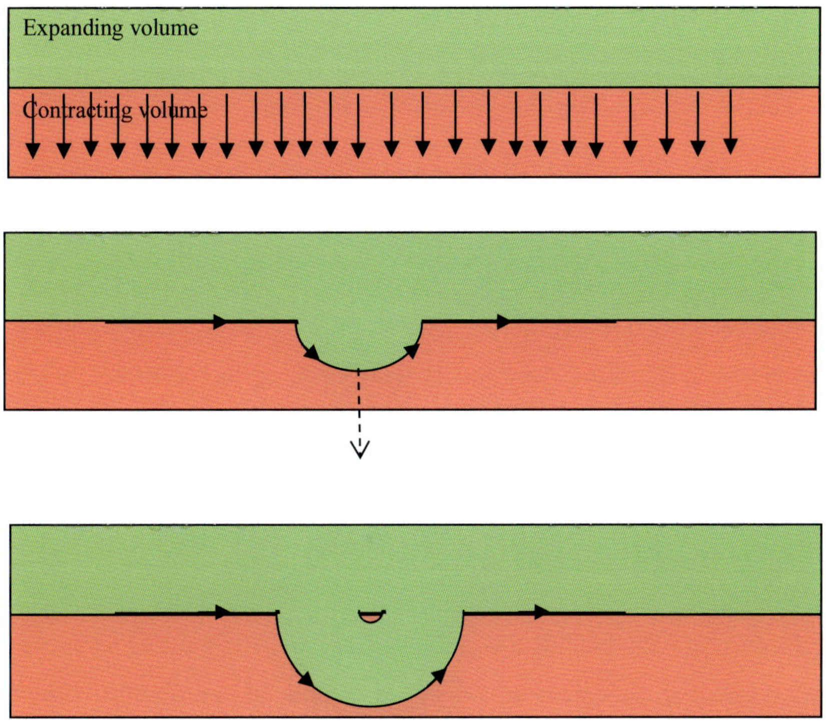

#D

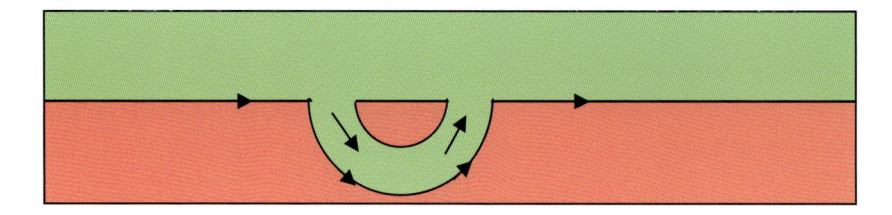

As massive amounts of vortices were formed, the density of the expanding space and the contracting space became equal, halting the initial creation.

It is theorized that during the initial creation, matter was packed so tightly together that matter anti-matter annihilations occurred. This created a tremendous amount of photons whose outward push into space gave them anti-gravity properties.

It is further theorized that this created an immense cloud of particles and photons. If the initial particles were Lambdas, the decay products would be protons & anti-protons, electrons & positrons, and neutrinos & anti-neutrinos.

At this point in time, it is theorized that this massive cloud of particles continued to expand outward as the volume of space they were imbedded in continued its outward expansion.

However, at the same time, there was no "unification of the forces of nature" as theorized by Contemporary Science. Since three dimensional space would be flowing into one end of the vortex and out of the other end, the charges of the particles would always be there from the very start; and, because of the rotation of particles, the two parts of the electromagnetic force (the electrostatic and electromagnetic forces) that these flowing microscopic rivers create, would be created with the 3d holes.

The force of gravity and anti-gravity were also created at the very beginning along with the particles. As explained in Books 2 & 3, the space flowing into a particle creates a field of less dense space that in turn creates "nuclear gravity" [explained in Book 3]. (The sum of all the nuclear gravity forces created around all of the protons and neutrons within a star are responsible for creating its gravitational field.) The Anti-gravity force would also be created at the very same time as matter because the hole at the other end of the vortex [an anti-proton, or an electron] would push space outward, creating the anti-gravity effect.

This would allow protons to attract electrons creating hydrogen atoms within the cloud; and anti-protons and positrons to also attract each other creating massive annihilations that created a simply fantastic amount of photons. It was indeed a "Big Bang"!!!

However, if the forces of nature were not initially negated in some way, particles and anti-particles would all be attracted to each other and annihilate. If this would have happened, no matter would be left in the universe. Hence, it is theorized that the only phenomenon that could stop this annihilation would be the massive amount of photons created during these early annihilations. As explained in Books 1, 2, & 3, photons push space outward, creating a denser region of space surrounding them, creating the anti-gravity effect. The anti-gravity effects of these photons - [that later became the CMB Cosmic Background Radiation] - crammed together into this massive though fantastically compacted space; would have tried to "push" all the forming matter away from each other: creating for lack of a better phrase, a "buffer zone" between matter and anti-matter.

The anti-gravity effect of the photons, combined with those constantly being absorbed and expelled by the particle anti-particle pairs would have created extreme agitations in this massive cloud of particles; which in turn would be responsible for creating enormous temperatures. These enormous temperatures, in turn would have created extremely fast motions within all of the matter in the cloud, and would have kept hydrogen and anti-hydrogen from forming until the initial formation of the particles of matter ceased, and the cloud began to expand.

These photons would also have kept all of the more massive particles from trying to condense into gigantic, massive nuclei that would have halted the ability of matter to later form the smaller nuclei needed for the individual atoms we see today.

The strong force [the exchange of quarks between protons and neutrons explained earlier] would slowly come into being when the nuclear gravity of protons attracted the nuclear gravity of neutrons creating the hydrogen nuclei of Deuterium and Tritium. The weak force would come into being last, when free neutrons decayed into protons, electrons and anti-neutrinos. Consequently, at no time were the forces of nature combined into "One" single force as previously proposed by Contemporary science.

THE CREATION OF HYDROGEN AND ANTI-HYDROGEN

When hydrogen and anti-hydrogen began to form, some of these atoms being in close proximity to each other would annihilate creating more photons; while other atoms of hydrogen and anti-hydrogen would push away from each other, creating "pockets" of matter, and anti-matter atoms. As these two groups expanded to form larger regions, the hydrogen atom regions eventually condensed and became the galaxies of the universe, while the anti-matter hydrogen regions could not coalesce and retreated into the vast regions of the seemingly empty space between the galaxies; eventually becoming what science calls the "Dark Energy" of the universe.

Note: as explained in Books 1, 2, & 3, because the photon is surrounded by a region of space bent outward, making it denser, creating anti-gravity effects; and the anti-matter proton is also surrounded by a region of space bent outward, making it denser, [creating its anti-gravity effects]; the anti-proton in the anti-hydrogen atom cannot absorb a photon. However, since the positron has a less dense region of space bent inward surrounding it, it is possible for an anti-hydrogen atom to absorb a photon and emit it through the anti-proton.

Hence, its wavelength emissions could still be calculated using the Rydberd Formula and its most easily seen spectral lines would be those of the Layman ultraviolet series, and the Balmer visual emission series. It is theorized that the infrared Paschen lines would not be as prevalent due to the fact that the anti-gravity effects of the anti-hydrogen, there would not be as many collisions between anti-hydrogen; and much less infra-red photons would be generated.

Later, as the space of the universe expanded outwards, the anti-gravity force created by the anti-hydrogen filling the spaces between the galaxies caused the galaxies to accelerate outward and away from each other. This caused the three dimensional volume of space in expansion to stretch outward, further accelerating the expansion of the universe. When the creation of the hydrogen and anti-hydrogen atoms cleared up the original gigantic opaque cloud of particles, it became transparent. Allowing the photons that were trapped in this gigantic cloud of particles to finally be released and travel outward into the universe becoming the Cosmic Microwave Radiation [widely referred to today by the three letter acronym: the CMB].

THE PROBLEM WITH THE COSMIC MICROWAVE BACKGROUND RADIATION?

Contemporary science states that the Cosmic Microwave Background Radiation occurred approximately 380,000 years after the "Big Bang". The scattering of this radiation by matter made it change direction from "Random Traveling" about between atoms and particles in the cloud, to move outward and away from the cloud in directions perpendicular to it. Supposedly, according to contemporary science, this "scattering" is the reason why we can still see it today. However, there is one big problem with this scenario: *why do we even see the background radiation at all?*

As hydrogen and helium began to form, the massive opaque cloud of matter began to turn transparent, allowing the release of the photons trapped within cloud. It is important to note that the release of these photons would have <u>preceded</u> the formation of the galaxies; consequently, the velocity of these photons [i.e. speed of light] should have sent them out into the space of the universe, never to be seen again! But this did not happen!

Note: the galaxies would have been moving slower than the speed of light, hence the faster moving photons would have moved past the galaxies ages ago.

Figure 20.2 The photons leave the cloud at the speed of light and travel out into the universe.

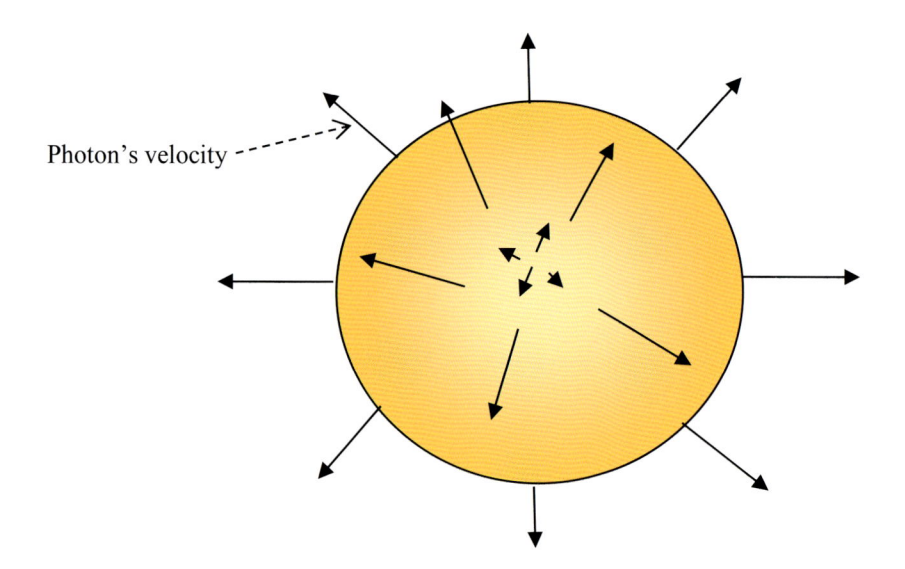

Note: the cloud is represented here in an idealistic spherical shape; the matter within the cloud cannot be seen.

According to contemporary science, the galaxies next form out of the cloud.

The photons are moving out into space at the speed of light "C" away from the galaxies. But look at their position in relation to the galaxies! According to this scenario, moving at a faster rate out into the universe, they have passed the galaxies and we should not be able to see them! But since we do, what is going on here!

Figure 20.3

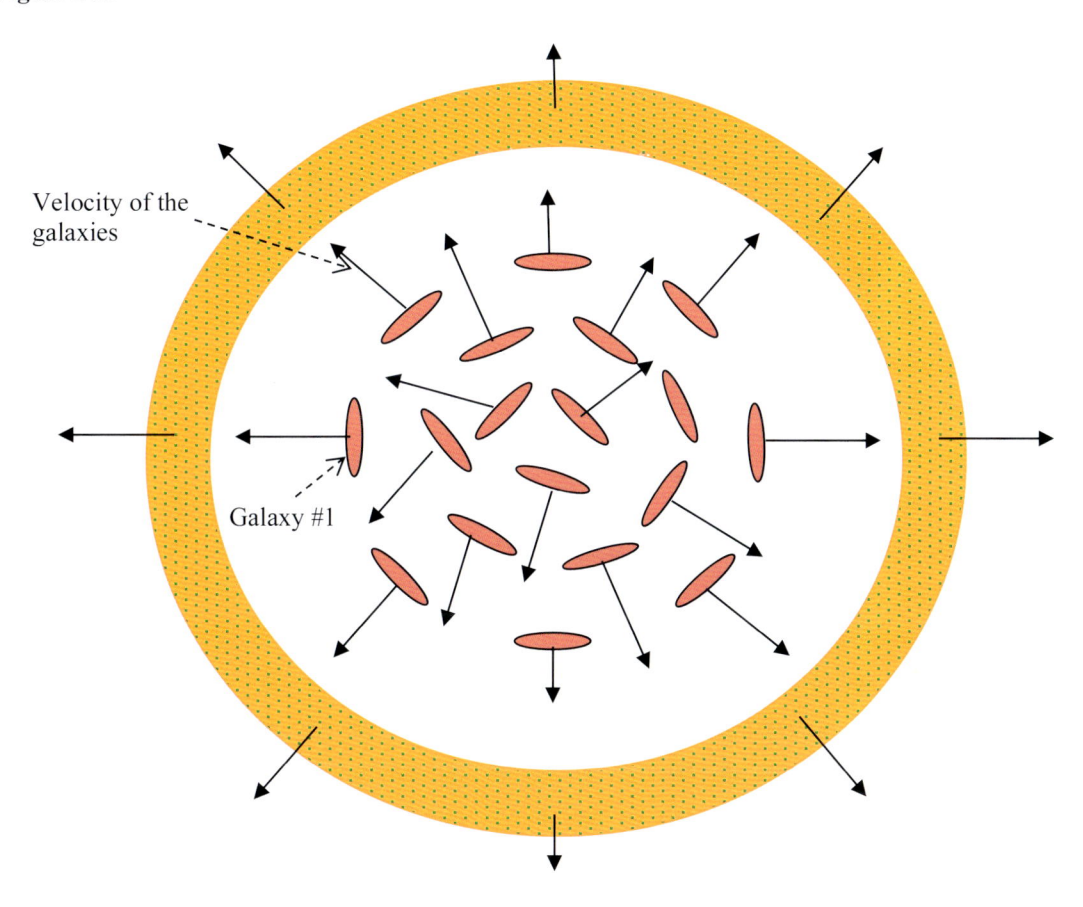

Again, according to the creation of the universe via contemporary science's big bang chain of events, the creation of the background radiation would have preceded the formation of the galaxies. When the opaque universe finally began to form hydrogen and turned transparent, the trapped photons of light would have gone out into the space beyond the galaxies of the universe as a giant halo of light never to be seen again! And here is where the problem develops:

BECAUSE WE CAN INDEED SEE THE CMB RADIATION, IT REVEALS A STUNNING, SHOCKING DISCOVERY ABOUT THE CONSTRUCTION OF OUR UNIVERSE!

Chapter 21
The Shape of the Universe Is the Solution to Seeing CMB Radiation

> The spherical shape of the universe is the solution to the return of the CMB radiation.

If the CMB radiation left the cloud before the creation of matter, it should have gone out into the universe never to be seen again. But we do see it! So how can such a seemingly impossible situation occur? Contemporary science cannot answer this problem. In fact, most have never even thought of this problem, or even considered the implications of this mystifying effect! Furthermore, there is no reference to this mysterious phenomenon in any of the cosmological literature on the ability to see this CMB radiation! Is there a solution to this mystery?

THE SOLUTION: THE SHAPE OF THE UNIVERSE...

The solution to this mystery is absolutely fascinating. It begins with the discovery that the volume of space of the universe we live in is not a volume of three dimensional space at all. Instead, it is the three dimensional spherical <u>surface</u> of a higher dimensional volume of space! Even though we do not know the exact shape of this surface, it can be theorized that it is symmetrical. If not, the intensity of the CMB light would vary from one parsec to another in the night sky.

Additional evidence of the shape of the universe can be found in the shape of the three dimensional holes the matter of the universe is made out of. Because the mathematics of the Vortex Theory Thesis revealed that the particles that make up matter are three dimensional spherical holes in the three dimensional surface of three dimensional space, we can further postulate that these holes take the shape of the three dimensional space that surrounds them.

If the three dimensional surface was for instance, in the shape of an oval, such a non symmetrical surface would create different tensions in the fabric of space, keeping the three dimensional holes from being symmetrical. This discrepancy would also be revealed in the shape of the electrostatic fields surrounding particles. They would not be symmetrical.

If all of the above is true, then, when the original cloud of particles was created, it would <u>not</u> take the asymmetrical shape envisioned below:

Figure 21.1

[Photons of light leaving the opaque cloud in straight lines as it slowly becomes transparent.]

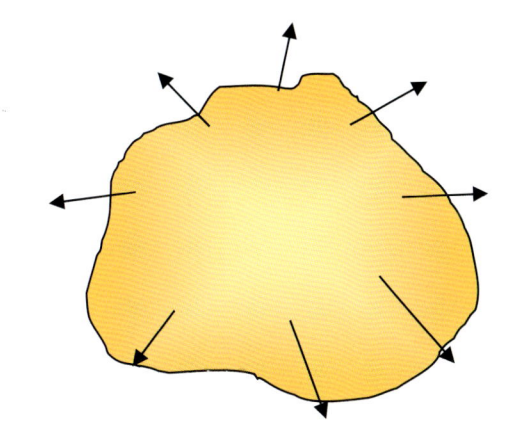

Instead it would be a thin cloud of particles upon a spherical surface seen below in the schematic of the cross section of the expansion of the three dimensional space of the universe seen below.

Figure 21.2

Cross section of surface of the universe: "an imaginary north pole" is at top:

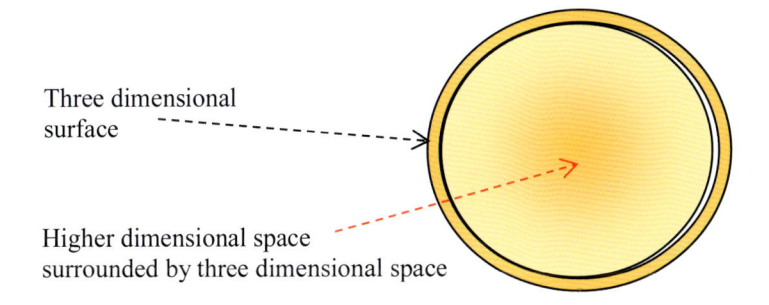

Three dimensional surface

Higher dimensional space surrounded by three dimensional space

It is also important to note that the photons of light will not travel outward in straight lines as assumed via the incorrect vision of the opaque cloud as seen in Figure 21.2. In reality, photons would travel in curved lines that merely appear straight from our point of view. This phenomenon is seen below on the cross section of the curved surface of the schematic of the universe:

Figure 21.3

In this schematic drawing of a two dimensional cross section of the three dimensional space of the Surface of the universe; notice how the photons can travel [from our point of view] in straight lines IN BOTH DIRECTIONS at once! Allowing any object upon the surface to see photons coming to it from two directions at once.

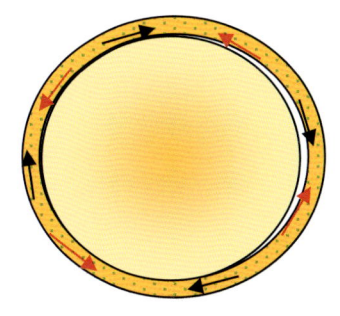

This revelation is most profound. Because it reveals that the CMB photons moving away from matter can travel all the way around the surface and arrive in all directions at once to any point in space such as our earth as seen in Figure 21.4 below!!!

Figure 21.4

Note: these two pictures represent an idealized schematic of the cross section of the universe; "cut in half" at an imaginary "equator" and viewed from a point above the "North Pole". The red oval in the center represents our Milky Way Galaxy with our Earth located inside. The arrows represent incoming CMB photons coming towards our Earth from all directions at once.

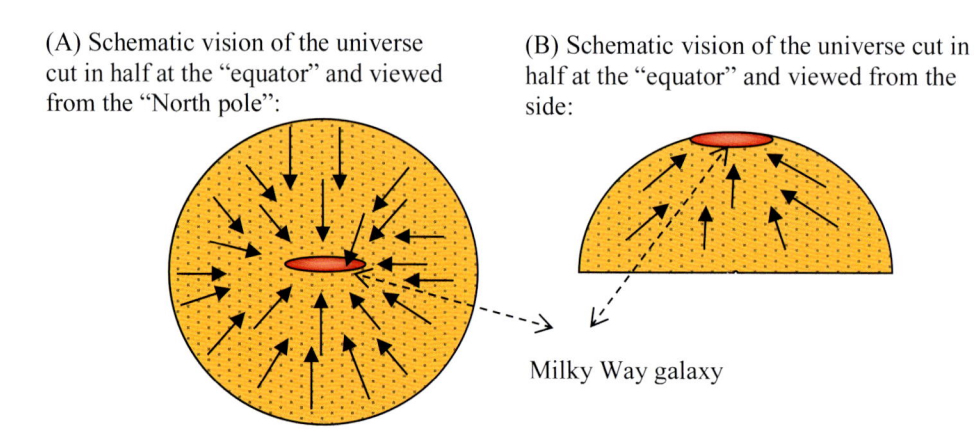

(A) Schematic vision of the universe cut in half at the "equator" and viewed from the "North pole":

(B) Schematic vision of the universe cut in half at the "equator" and viewed from the side:

Milky Way galaxy

Chapter 22
A Problem With the Red Shift!

> Is there a problem with the Red Shift of the most distant galaxies?

Edwin Hubble was one of the world's greatest astronomers. His discovery of the Red Shift of the galaxies was one of his greatest achievements. His use of it to determine the age of the universe also appears to be a great discovery. However, like many other astronomers of the past and present, he made one fatal mistake; he did not understand the physical construction of space and the shape of the universe.

Like everyone else, Hubble assumed that when we look out into the universe with a telescope, that we are looking out into a volume of space. Furthermore, he accepted Einstein's proposal and assumed that space is made of nothing; hence, it cannot influence the velocity of the galaxies. But this is not true.

One of the great discoveries of the Vortex Theory Atomic Particles reveals that the universe is one single gigantic particle of immense size whose three dimensional surface is in expansion. On a small scale, the force of gravity causes galaxies to move towards each other to form small clusters such as our "Local Group", and hold together mighty galactic structures such as the "Wall". But on a large scale, the great red shifts of the galaxies are caused by the expanding space of the three dimensional surface of the universe.

Figure 22.1

In the Figures below, notice how the expanding radius of the circle makes the circumference grow larger, increasing the distance between galaxies A & B.

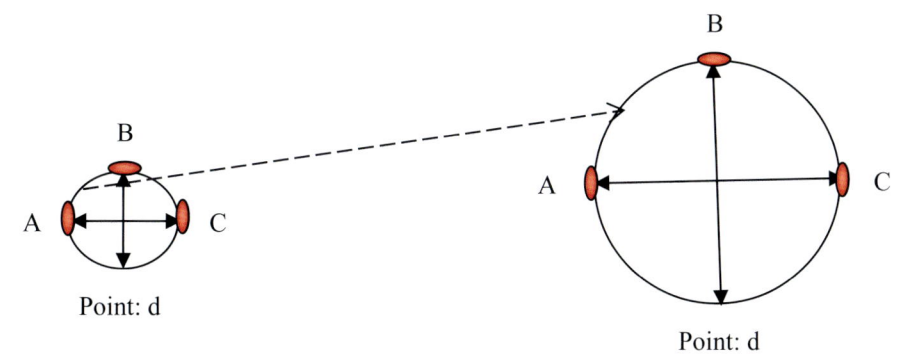

Point: d

Point: d

This expanding space, increases the distance between galaxies, making the light traveling from one galaxy to another "Red Shifted"; making it seem from our point of view as if the galaxies are racing away from each other.

Another fascinating mystery is the light coming from galaxy C traveling along the path from C to B to A; and from C to d to A? Does this mean we can see the galaxy at B twice? Do we see it when looking in one direction, and then see it again when we turn our telescope 180^0 in the opposite direction and see it again? This phenomenon will be discussed at the end of this book in an experiment that can be performed by astrophysicists.

Chapter 23
The Secret of the Universe

> The true age of the universe is not determined by the greatest red shifted galaxy we can see. Nor will it be determined by the greatest red shift of a galaxy we can see in both directions 180 degrees in reverse to each other. It will have to be determined by the formation of Population II stars containing small amounts of heavy metals.

Using the principles of T*he Vortex Theory of Atomic Particles*, over 80 great and wonderful scientific discoveries have been made. But none of them can compare to the one that stands upon the apex of all the others; by far, the greatest of them all: something that can only be called *THE SECRET OF THE UNIVERSE*! The discovery that the massive volume of space we live in – the three dimensional volume that stretches from galaxy to galaxy, across the length and breadth of the entire universe – is the surface of one single gigantic particle!

Even more astonishing, because this Vortex Theory reveals that all of the tiny sub-atomic particles matter is constructed out of are actually three dimensional holes imbedded within the surface of this one gargantuan particle; then, all of the matter in the universe is "one" with each other! Shockingly, we are not the separate entities we perceive ourselves to be; nor are the planets, suns, and galaxies. Instead, we and everything else that exists in the universe are all part of this one same "object"! It is almost unbelievable to contemplate, but we are as much a part of all the stars, planets, and galaxies as they are with us!

Suddenly the ancient statements of mystics and sages of old begin to make sense. Statements such as, "Oneness"; "All is One"; and, "I am that, thou art that, that is that" take on a new meaning. Because in fact we are all "One", we are all a part of the whole. The idea of ""Oneness" suddenly possesses profound implications. All of the matter in the universe is unexpectedly equivalent to a massive collection of corpuscles existing within the colossal body of one object! Nor is this speculation or supposition; this fantastic new, and revolutionary vision of the universe and ourselves is based upon the scientific evidence uncovered during 30 years of investigations using *The Vortex Theory of Atomic Particles*.

Even more important, there is a way to prove it with the following experiment!

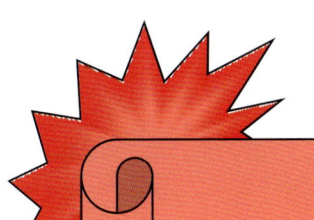

Chapter 24
An Experiment for Astronomers & Astrophysicists!

Is the universe one gigantic individual particle turning inside out as proposed by this Vortex Theory of Atomic Particles? Is there a way of proving it? The answer is yes!

The following experiment is proposed for astronomers and astrophysicists. It is easily done, and most important of all, costs no money!

The proposal by the Vortex Theory of Atomic Particles, that our ability to see the CMB radiation created at the beginning of the universe is a function of the curved surface of the universe…has profound implications! For it also reveals that the starlight we see coming to us from the most distant galaxies with the greatest red shifts is not coming to us in straight lines such as that from the close by Andromeda galaxy; but rather, from all the way around the curvature of the other side of the universe! Even more shocking, some of it might be coming from around the "back side" of the universe, circling all the way around, to finally reach us after traveling trillions and trillions of miles. For example, note the figure below:

Figure 24.1

In the figure below, representing a cross-section of the universe of diameter d, there are six galaxies: A, B, C, D, E & F. A represents the Milky Way Galaxy; B represents the Andromeda Galaxy; while C, D, E, & F represents galaxies existing upon the three dimensional surface of the expanding higher dimensional particle with large red shifts.

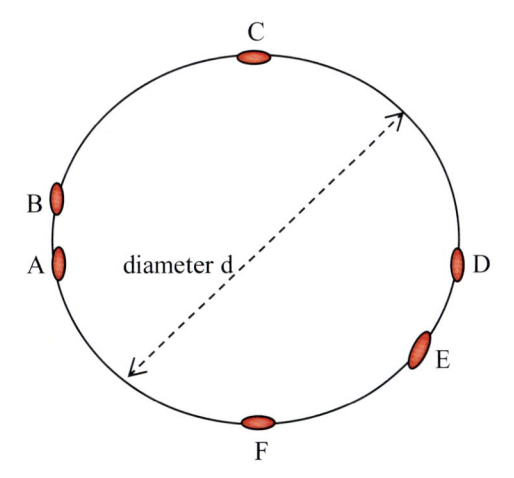

Looking at Figure 24.1 we can see that light flowing from B to A, moves for all practical purposes in a straight line. However, light flowing from C to A has to move around the curvature of the particle. The same is true for D. D can travel in <u>two</u> paths to A: it can go from C to B and then to A; or it can go from E to F to A.

The same is true for E: E can go from D to C to B to A; or, to F, and then to A.

But what is most fascinating is the fact that the light from any of these galaxies is traveling both clockwise towards us and counter clockwise towards us at the same time! In effect, depending upon the radius R, and the age of the galaxy in question, we can be seeing double!

Galaxy D presents an even more shocking illusion. Although Figure 24.1 only represents a two dimensional slice of the universe, in reality it is a sphere. Consequently, the light from Galaxy D can be coming to us from all directions at once! In other words, in any direction we point our telescope we should be able to see it! However, this hypothesis is based upon the amount of starlight we can see. If it is bright, we should be able to see it in any direction making it appear as if there are hundreds of galaxies instead of only seeing one! Then again, if the light coming from Galaxy D is weak, we might not even be able to see it at all!

The curvature of the three dimensional surface of the universe suddenly creates a big problem for astronomy. Instead of seeing Billions of galaxies we might be only seeing one half or one fourth of what actually exists! Can this dilemma be resolved? The answer is yes.

There is an experiment that can be conducted to resolve this problem. Here is how it works…

We already have extensive star maps of the night sky. Although many of the distant galaxies are just points of light on these maps, never the less, we can still see them. Therefore, because there are so many galaxies, all of the pictures of every parsec of the universe can be fed into a computer.

Next, the computer can be programmed to look at all of the small galaxy clusters noting their Azimuth and Altitude. Such as that seen below:

Figure 24.2

In the figure below, the computer sees a small star cluster, then immediately looks in the opposite Altitude and Azimuth to see if it can see it again.

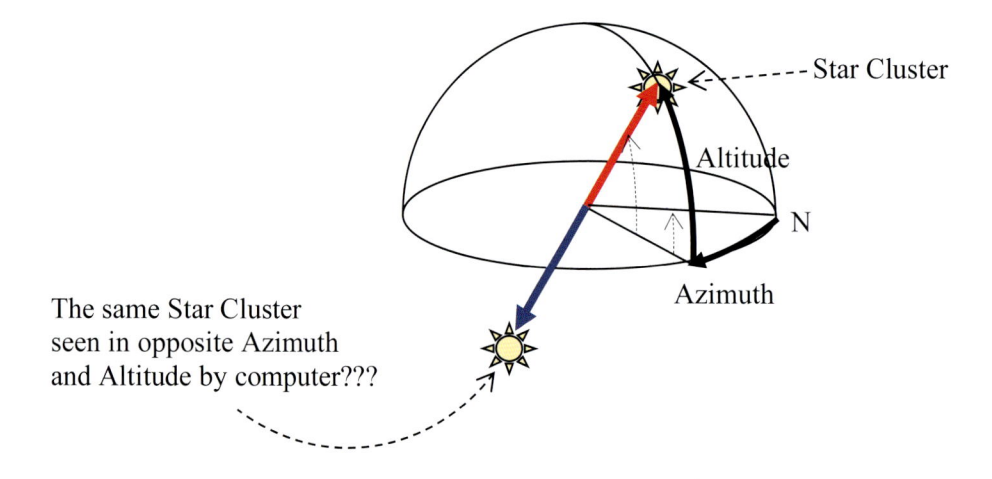

If the same cluster is seen, then, if unknown, the red shifts of both clusters must be determined by an actual telescope observation to see if they are the same.

It is proposed to look for clusters first rather than individual galaxies because it is easier to identify clusters with an equal amount of galaxies.

One of the drawbacks in looking at clusters comes from the realization that if the red shifts of the opposite clusters are not the same, individual movements of the galaxies within the clusters might not make them readily identifiable to the computer. This is revealed by the moments of the galaxies within our own cluster known as the Local group. In our Local Group, the Andromeda Galaxy is moving towards us and should reach us in about 4 billion years. Consequently if we are looking at a cluster of galaxies where some are about to collide, then from the opposite direction and if the red shift is not the same, some of the galaxies could have collided and the computer will no longer recognize them as the same cluster we see in the opposite direction.

Looking at Figure 24.1 again, and renaming it Figure 24.3 allows us to make additional observations that if made before would be considered premature.

Again, the figure below, representing a cross-section of the universe of diameter d, there are six galaxies: A, B, C, D, E & F.

Figure 24.3

A represents the Milky Way Galaxy; B represents the Andromeda Galaxy; while C, D, E, & F represents galaxies existing upon the three dimensional surface of the expanding higher dimensional particle with large red shifts.

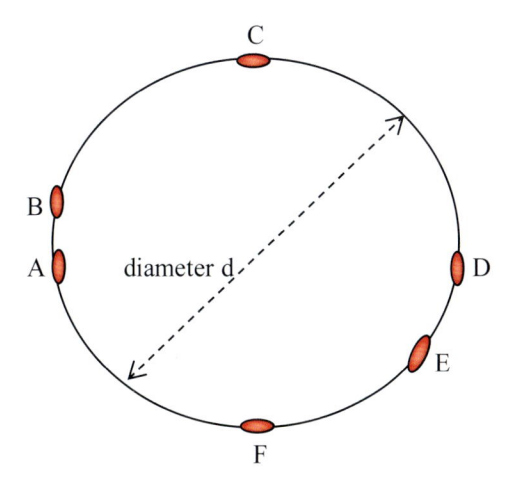

If light traveling the entire circumference of the particle takes what appears to be a red shift of 2, then, the light traveling half way around from D to A, will have a red shift of 1. And we can then see that no matter what direction the light from D takes to reach us, we will always see a red shift of 1 for this galaxy [we will come back to this most important galaxy shortly].

However, the light coming from Galaxy E has two different red shifts. Say for example when the light travels from D to C to B to A, it is seen to have a red shift of 1.2. Then when the light travels the path past F to reach A, it will have a red shift of 2 – 1.2 = .8! This is a very important observation because it reveals that the light from one galaxy but seen in the opposite direction appears to be a different galaxy of different age!

This observation makes it seem as if there are many more galaxies in the universe!

Returning our attention to galaxy D, we can call this the "Holy Grail" of Galaxies. This label is given to it because again, as observed above, we will see it no matter what direction we look in! If it exists, and the probability is great that it does, if not, the reason why we do not see it, is that the light coming from it has to be very faint due to the fact that it has to be very far away.

We can now define the objectives of our experiment as:

#1 Find the same galaxy clusters in the exact opposite directions to each other.

#2 Next, find individual galaxies in opposite directions to each other.

#3 And finally, find the galaxy identified as the Holy Grail of all the galaxies in the universe.

   [Note, perhaps it is a cluster?]

The parameters of our search are only limited by the amount of galaxies our best telescope can see: the Hubble perhaps?

Chapter 25
Determining the True Age of the Universe?

> As mistakenly believed by contemporary science, the true age of the universe is not determined by the greatest red shifted galaxy we can see. Nor will it be determined by the greatest red shift of a galaxy we can see in both directions 180 degrees in reverse to each other [in the experiment proposed at the end of this book]. It will have to be determined by the formation of Population II stars containing small amounts of heavy metals.

THE AGE OF THE UNIVERSE

We have now seen that a galaxy with a great Red Shift might be an illusion. Its Red Shift might be caused by its light traveling to us from around the "backside" of the universe. Consequently, it can no longer be used to date the universe. The accelerating space of the universe has also created a problem for using the "Holy Grail" galaxy to date the universe.

In 1998, two independent projects, the Supernova Cosmology Project and the High-Z Supernova Search Team, discovered the expanse of the universe is accelerating. This eliminates the red shift of the "Holy Grail" Galaxy as the explanation for the age of the universe. If its velocity was steady we could trace its origin back to the beginning. Since it is accelerating, we cannot.

The only way we really have now to explain how long it took from the time of the "Big Bang" until the present is the formation of Population II stars containing large amounts of Hydrogen and Helium and small amounts of heavy metals. However, even this form of time dating criteria has its problems.

The formation of Population II stars seems to reveal that the age of the universe is approximately 13.66 billion years old, [older than that proposed via the red shift]. This comes from the fact that the star identified as HD 140283 appears to be 13.66 billion years old. This discovery made *by* H. E. Bond; E. P. Nelan; D. A. VandenBerg; G. H. Schaefer; and D. Harmer in 2013 seems to exhibit the best evidence. However, out of all of the hundreds of billions of stars in the universe, there's a good chance some star in some distant galaxy might be older.

Conclusion

> There are too many problems with the proposal of the Singularity being the cause for the creation of the universe. The creation scenario proposed by the Vortex Theory of Atomic Particles explains all of the phenomenon of the universe the Singularity cannot. From the Big Bang to quarks, it does it all.

There are too many problems with Conventional Science's explanation for the creation of the universe. It is predicated upon pure speculation of the existence of a mythical particle called the Singularity. Nobody can explain where it came from! Neither can anyone explain how all of the matter in the universe - a simply gigantic, massive, unimaginable amount of matter contained within a hundred billion suns in a hundred billion galaxies - could be contained within this one single micro-dot! Nor is there any precedent in nature to base the assumption of the Singularity compressing matter into a smaller and smaller volume.

Note, even black holes, some containing the volume of billions of suns have a large measurable diameter. And this is a problem. Using the logic of the singularity, the more material a black hole contains the smaller it should get. These massive collections of matter should be getting smaller and smaller, approaching the size of the original "micro-dot" of matter! Unfortunately for the theory of the singularity, this is not happening.

Instead, the black hole designated M60, has a diameter of 45 billion miles, eight times the diameter of Neptune's orbit around the Sun; and it possesses 4.5 billion suns in its massive volume.

Neither can anyone explain *how or why* the singularity exploded causing the Big Bang? Why did it blow up? Nobody knows? And even more important, where is the point of origin of the blast? All explosions have a point of origin. As mentioned before, the supernova that created the "Crab Nebula", and documented by the Chinese in 1054 AD has a point of origin whose location can be extrapolated from analyzing the Doppler shift of the matter racing out into space in all directions at once. If the singularity had existed and exploded, all of the galaxies of the universe should all be rushing away from *one single point of origin*! But this is not what we see. Our telescopes cannot locate a point of origin.

Next comes the proposal that there was a time when there existed a massive cloud of quarks and gluons. And yet, in all of the particle collisions ever conducted in all of the linear accelerators in the world, and amongst all of the trillions and trillions of particles created in these collisions, no individual quark has ever been seen!

Photons and neutrinos are also keys to understanding how the universe is constructed. Photons are condensations of three dimensional space; while neutrinos are quantized transverse waves [similar to tennis balls cut in half] on the two different surfaces of the two volumes of space: one in contraction one in expansion.

It should also be noted, that the advocates of the String Theory think they can explain the singularity if all matter is made out of one dimensional vibrating strings. Their logic states that if this is so, then all the one dimensional strings in the universe can blend together into one particle with no three dimensional thickness. However, if time does not exist which the mathematics of the Vortex Theory Thesis reveals, then neither does Einstein's fourth dimension of spacetime:

[The Thesis and for which the Russian Ministry of Education awarded a PhD in 2005 (note: in Russia all higher degrees are awarded by the Russian Ministry of Education); and whose PROOF was published by the St. Petersburg State University's Branch of the Russian Academy of Sciences in 2012]. Consequently, if the fourth dimension of spacetime doesn't exist, the String Theory is INCAPABLE of explaining the length shrinkage and time dilation effects associated with near light velocities. And as these theorists well know, ANY theory that attempts to explain the construction of the universe HAS TO EXPLAIN length shrinkage and time dilation effects: and if it can't, it is Defunct!]

Shockingly, and again, as explained in Books 2 & 3, every point in the universe also appears to be the center of the universe! This is a phenomenon created by a surface and not a volume. For example, from the perspective of the volume of space within a basketball, there is only one point that is the center of the volume. However, from the perspective of a tiny ant standing upon the outside surface of the basketball, as he slowly turns around and looks in all directions as far as he can see, he appears to be standing upon the center. And even if he shifts his location to any other place on the surface of the ball, he continues to see the same phenomenon: making it appear as if every point on the surface of the sphere is the center.

Then comes the assumption about force. Contemporary science proposes that all the forces of nature are manifestations of one single force created during the Big Bang! But again, this is pure speculation. Albert Einstein spent his life trying to prove this assumption and failed. As explained in Books 1, 2, & 3, it is revealed that the forces of nature are created out of denser and less dense space; bent and flowing space; with the exception of the strong force, whose construction is created by the exchange of a virtual Pion between protons and neutrons. Hence, they are not all manifestations of "one" force.

And finally, comes the problem with CMB radiation: why is it coming to us from all directions at once? It should have passed by us billions of years ago, never to be seen again! In contemporary science's explanation of the creation of the universe, the photons of light trapped within the original "gigantic plasma cloud" in the early universe, were released into the universe before the creation of the matter that subsequently condensed into the galaxies. So how come we can still see it? Like a flashlight shining its light out into the universe, once the light is on its way - it is gone, never to be seen by us again! However, since we can indeed see the CMB radiation, it reveals that the true shape of the universe is not that which present cosmologists propose, but instead, is a shocking vision unlike anything anyone has ever imagined before; and explained in all three Vortex Theory books along with all of the other former mysteries of science.

Then there is the work of Sir Roger Penrose, whose investigations into the abnormalities seen in the CMB radiation, seems to reveal the universe was created and destroyed before. This is a confirmation of the Vortex Theory of Atomic Particle's earlier, though discounted conclusion, that the explanations for the diverse structures of the galaxies of the universe [such as the Wall, the Strings of galaxies, and the large Clusters of galaxies] are a result of a cyclic universe. An oscillating universe created and destroyed many times before, and not a single "once in a lifetime" event.

The answers to all of these former mysteries of science are finally solved by the vision of creation presented in the *Vortex Theory of Atomic Particles*. In this Theory, it is revealed that the three dimensional space of the universe is not a volume, but instead, is the surface of a higher dimensional object. That the invisible space of this object is made of something and everything in the universe is created out of it: particles of matter are three dimensional holes embedded within the three dimensional space of its surface; the forces of nature are made out of bent and flowing space, or created by the movement of virtual particles from one larger particle to another. Photons are condensations of three dimensional space; while neutrinos are quantized transverse waves [similar to tennis balls cut in half] on the two different surfaces of the two volumes of space: one in contraction one in expansion. There is no other one theory of the universe that can explain all of this. Nor is there any other theory that explains all of the other great mysteries of physics. Because of this, it becomes clear that the Vortex Theory of Atomic Particles is the only complete and correct explanation for the creation of the universe.

References

National/International Conferences attended, and peer reviewed scientific papers presented

[1] The Vortex Theory of Matter. [Presentation of his own work]
'International Forum on New Science' Colorado State University (1992, Sept 17-20). Moon. R. Fort Collins, Colorado. USA. Topic: The Vortex Theory of Matter. Copyright 1990)

[2] The Bases of the Vortex Theory. [Paper presentation]
'The LIII International Meeting on Nuclear Structure; Nucleus 2003' (2003, October 7-10). Moon, R. Vasiliev, V. p. 251. Saint Petersburg, Russia.

[3] The Vortex Theory and some interactions in Nuclear Physics. [Book of abstracts; p. 259]
'The LIV International Meeting on Nuclear Spectroscopy and Nuclear Structure; Nucleus 2004' (2004, June 22-25). Moon, R., Vasiliev, V. Belgorod, Russia.
http://nuclpc1.phys.spbu.ru/nucl/Abstracts/Nucleus_2004.pdf

[4] The Possible Existence of a new particle: The Neutral Pentaquark. [Book of materials; pp. 98-104]
'Scientific Seminar of Ecology and Space' (2005, February 22). Scientific Research Centre for Ecological Safety of the Russian Academy of Sciences. Moon, R. Saint Petersburg, Russia.
https://spcras.ru/ensrcesras/

[5] Explanation of Conservation of Lepton Number. [Book of materials; p. 105]
'Scientific Seminar of Ecology and Space' (2005, February 22). Scientific Research Centre for Ecological Safety of the Russian Academy of Sciences: Moon, R., Vasiliev, V. Saint Petersburg, Russia.
https://spcras.ru/en/srcesras/

[6] Explanation of Conservation of Lepton Number. [Book of abstracts; p. 347]
'LV National Conference on Nuclear Physics' (2005, June 28-July 1). FRONTIERS IN THE PHYSICS OF NUCLEUS. Moon, R., Vasiliev, V. Russian Academy of Sciences. St. Petersburg State University. Saint Petersburg, Russia.
http://nuclpc1.phys.spbu.ru/nucl/Abstracts/Frontiers_2005.pdf

[7] The Possible Existence of a New Particle: the Tunneling Pion. [Book of abstracts; p. 348]
'LV National Conference on Nuclear Physics' (2005, June 28-July 1). FRONTIERS IN THE PHYSICS OF NUCLEUS. Moon, R., Vasiliev, V. Russian Academy of Sciences. St. Petersburg State University. Saint Petersburg, Russia.
http://nuclpc1.phys.spbu.ru/nucl/Abstracts/Frontiers_2005.pdf

[8] The Possible Existence of a New Particle in Nature: the Neutral Pentaquark. [Book of abstracts; p. 349] 'LV National Conference on Nuclear Physics' (2005, June 28-July 1). FRONTIERS IN THE PHYSICS OF NUCLEUS. Vasiliev, V. Moon, R. Russian Academy of Sciences. St. Petersburg State University. Saint Petersburg, Russia.
http://nuclpc1.phys.spbu.ru/nucl/Abstracts/Frontiers_2005.pdf

[9] The Experiment that discovered the Photon Acceleration Effect. [Book of abstracts; p. 77]
'International Symposium on Origin of Matter and the Evolution of Galaxies' (2005, Nov 8-11). Gridnev, K., Moon, R., Vasiliev, V. New Horizon of Nuclear Astrophysics and Cosmology. University of Tokyo, Japan.
https://meetings.aps.org/Meeting/SES05/Content/273
https://flux.aps.org/meetings/bapsfiles/ses05_program.pdf

[10] The Conservation of Lepton Number. [Paper presentation]
'American Physical Society 72nd Annual Meeting of the Southeastern Section of the APS' (2005, Nov 10-12). Moon, R., Calvo, F., Vasiliev, V. Gainesville, FL. USA. APS Session BC Theoretical Physics I, BC 0008
https://meetings.aps.org/Meeting/SES05/Content/273
https://flux.aps.org/meetings/bapsfiles/ses05_program.pdf

[11] The Vortex Theory and the Photon Acceleration Effect. [Paper presentation]
'American Physical Society; March Meeting; Topics in Quantum Foundations' (2006, March 13-17). Gridnev, K., Moon, R., Vasiliev, V. Baltimore, Maryland. USA.
Abstract ID: BAPS.2006.Mar.B40.6
https://meetings.aps.org/Meeting/MAR06/Session/B40.6
http://meetings.aps.org/link/BAPS.2006.MAR.B40.6

[12] The St Petersburg State University experiment that discovered the Photon Acceleration Effect.
'American Physical Society; March Meeting' GENERAL POSTER SESSION (2006, March 13-17).Gridnev, K., Moon, R., Vasiliev, V. Baltimore, Maryland. USA.
Abstract ID: BAPS.2006.MAR.Q1.146
https://meetings.aps.org/Meeting/MAR06/Session/Q1.146
http://meetings.aps.org/link/BAPS.2006.MAR.Q1.146

[13] The Neutral Pentaquark.
'American Physical Society; March Meeting' GENERAL POSTER SESSION (2006, March 13-17).Moon, R., Calvo, F., Vasiliev, V. Baltimore, Maryland. USA.
Abstract ID: BAPS.2006.MAR.Q1.147
https://meetings.aps.org/Meeting/MAR06/Session/Q1.147
http://meetings.aps.org/link/BAPS.2006.MAR.Q1.147

[14] The Neutral Pentaquark. [Paper presentation]
'International Workshop on "Nuclear Physics with RIBF' (2006, March 13-17).
Vasiliev, V., Calvo, F., Moon, R. RIKEN Research Institution. Saitama, JAPAN.
Abstract: RIBF-Pentaquark.
https://ribf.riken.jp/RIBF2006/

[15] Nuclear Structure and the Vortex Theory. [Paper presentation]
'International Workshop on "Nuclear Physics with RIBF' (2006, March 13-17).
Moon, R., Vasiliev, V. R. RIKEN Research Institution. Saitama, JAPAN.
Abstract RIBF-Vortex
https://ribf.riken.jp/RIBF2006/

[16] Experiment that Discovered the Photon Acceleration Effect. [Paper presentation]
'International Workshop on "Nuclear Physics with RIBF' (2006, March 13-17).
Moon, R., Vasiliev, V. R. RIKEN Research Institution. Saitama, JAPAN.
Abstract Moon 1
https://ribf.riken.jp/RIBF2006/

[17] To the Photon Acceleration Effect. [Paper presentation]
'APS/AAPT/SPS Joint Spring Meeting' (2006, March 21-23).
Moon, R. San Angelo, Texas. USA. Abstract ID: BAPS.2006.TSS.POS.8
https://meetings.aps.org/Meeting/TSS06/Session/POS.8
http://meetings.aps.org/link/BAPS.2006.TSS.POS.8

[18] The Saint Petersburg State University Experiment that discovered the Photon Acceleration Effect. [Paper presentation] 'American Physical Society; Astroparticle Physics II' (2006, April 22-25).
Gridnev, K., Moon, R., Vasiliev, V. Dallas, TX. USA. Abstract ID: BAPS.2006.APR.J7.6
https://meetings.aps.org/Meeting/APR06/Session/J7.6
http://meetings.aps.org/link/BAPS.2006.APR.J7.6

[19] The Photon Acceleration Effect. [Paper presentation]
'American Physical Society; Session W9 DNP: Nuclear Theory II' (2006, April 22-25).
Gridnev, K., Moon, R., Vasiliev, V. Dallas, TX. USA. Abstract ID: BAPS.2006.APR.W9.6
https://meetings.aps.org/Meeting/APR06/Session/W9.6
http://meetings.aps.org/link/BAPS.2006.APR.W9.6

[20] The Neutral Pentaquark. [Paper presentation]
'American Physical Society; Session W9 DNP: Nuclear Theory II' (2006, April 22-25).
Moon, R., Calvo, F., Vasiliev, V. Dallas, Texas. USA. Abstract ID: BAPS.2006.APR.W9.9
https://meetings.aps.org/Meeting/APR06/Session/W9.9
http://meetings.aps.org/link/BAPS.2006.APR.W9.9

[21] Controversy surrounding the Experiment conducted to prove the Vortex Theory. [Paper presentation] 'American Physical Society; 8th Annual Meeting of the Northwest Section' (2006, May 18-20). Vasiliev, V., Moon, R. University of Puget Sound. Tacoma, Washington. USA.
Abstract ID: BAPS.2006.NWS.C1.9
https://meetings.aps.org/Meeting/NWS06/Content/518
https://meetings.aps.org/Meeting/NWS06/Session/C1.9

[22] The Photon Acceleration Effect. [Paper presentation]
'American Physical Society; 8th Annual Meeting of the Northwest Section' (2006, May 18-20). Moon, R., Vasiliev, V. University of Puget Sound. Tacoma, Washington. USA.
Abstract ID: BAPS.2006.NWS.C1.8
https://meetings.aps.org/Meeting/NWS06/Content/518
https://meetings.aps.org/Meeting/NWS06/Session/C1.8
http://meetings.aps.org/link/BAPS.2006.NWS.C1.8

[23] Experiment that Discovered the Photon Acceleration Effect. [Paper presentation]
'International Congress on Advances in Nuclear Power Plants' ICAPP '06, (2006, June 4-8).
Gridnev, K., Moon, R. Reno, Nevada. USA. American Nuclear Society.
Abstract 6006. ISBN: 978-0-89448-698-2

[24] The Neutral Pentaquark. [Paper presentation]
'International Congress on Advances in Nuclear Power Plants' ICAPP '06 (2006, June 4-8).
Vasiliev, V., Calvo, F., Moon, R. Reno, Nevada. USA. American Nuclear Society.
Abstract 6045. ISBN: 978-0-89448-698-2

[25] Is Hideki Yukawa's explanation of the strong force correct?
'The International Symposium on Exotic Nuclei' Book of abstracts: Joint Institute for Nuclear Research. (2006, July 17-22). Vasiliev, V., Moon, R. Khanty Mansiysk, Siberia. Russia.
http://wwwinfo.jinr.ru/exon2006/
http://jinr.ru/

[26] The Explanation of the Pauli Exclusion Principle. [Paper presentation]
'59th Annual meeting of the American Physical Society Division of Fluid Dynamics' (2006, Nov 19-21). Moon, R., Vasiliev, V. Tampa, Florida. USA. American Physical Society;
Abstract ID: BAPS.2006.DFD.P1.17
https://meetings.aps.org/Meeting/DFD06/Content/578
https://meetings.aps.org/Meeting/DFD06/Session/P1.17
http://meetings.aps.org/link/BAPS.2006.DFD.P1.17

[27] Is Hideki Yukawa's explanation of the strong force correct? [Paper presentation]
'59th Annual meeting of the American Physical Society Division of Fluid Dynamics' (2006, Nov 19-21). Moon, R., Vasiliev, V. Tampa, Florida. USA. American Physical Society;
Abstract ID: BAPS.2006.DFD.P19
https://meetings.aps.org/Meeting/DFD06/Content/578
https://meetings.aps.org/Meeting/DFD06/Session/P1.19
http://meetings.aps.org/link/BAPS.2006.DFD.P1.19

[28] The Final Proof of the Michelson Morley Experiment; The explanation of Length Shrinkage and Time Dilation. [Book of materials] 'Scientific Research Center for Ecological Safety of the Russian Academy of Sciences: Scientific Seminar of Ecology and Space'. (2007, February 8-10). Moon, R. Saint Petersburg, Russia.
https://spcras.ru/en/srcesras/

[29] The Explanation of the Photon's Electric and Magnetic fields and its Particle and Wave Characteristics. [Paper presentation] 'Annual Meeting of the Division of Nuclear Physics Volume 52, Number 10'. (2007, Oct 10-13). Moon, R., Vasiliev, V. Newport News, Virginia. USA. American Physical Society; Abstract ID: BAPS.2007.DNP.BF.15
https://meetings.aps.org/Meeting/DNP07/Session/BF.15
http://meetings.aps.org/Meeting/DNP07
http://meetings.aps.org/link/BAPS.2007.DNP.BF.15

[30] The St. Petersburg State University experiment that discovered the Photon Acceleration Effect. 'Virtual Conference on Nanoscale Science and Technology' VC-NST. (2007, Oct 21-25). Moon, R., Vasiliev, V. University of Arkansas. 222 Physics Building. Fayetteville, AR 72701 USA.
http://www.ibiblio.org/oahost/nst/index.html

[31] The Explanation of Quantum Teleportation and Entanglement Swapping. [Paper presentation] '49th Annual Meeting of the Division of Plasma Physics, Volume 52, Number 11' (2007, Nov 12–16). Moon, R., Vasiliev, V. Orlando, Florida. American Physical Society;
Abstract ID: BAPS.2007.DPP.UP8.21
https://meetings.aps.org/Meeting/DPP07/Content/901
http://meetings.aps.org/link/BAPS.2007.DPP.UP8.21
https://meetings.aps.org/Meeting/DPP07/Session/UP8.21

[32] The Explanation of the Photon's electric and magnetic fields, and its particle and wave characteristics. [Paper presentation]
'60th Annual Meeting of the Division of Fluid Dynamics'. Volume 52, Number 12. (2007, Nov 18–20). Moon, R., Vasiliev, V. Salt Lake City, Utah. American Physical Society;
Abstract ID: BAPS.2007.DFD.JU.22
http://meetings.aps.org/Meeting/DFD07
https://meetings.aps.org/Meeting/DFD07/Session/JU.22
http://meetings.aps.org/link/BAPS.2007.DFD.JU.22

[33] The Explanation of quantum entanglement and entanglement swapping. [Poster Session] 'The 10th International Symposium on the Origin of Matter and the Evolution of the Galaxies (OMEG07) (2007, Dec 4-6) Moon, R., Vasiliev, V. Hokkaido University, Sapporo, Japan. Bibcode: 2008AIPC.1016.....S, Harvard (Astrophysics Data System) ISBN 0735405379
https://ui.adsabs.harvard.edu/abs/2008AIPC.1016.....S/abstract

[34] The Explanation of the Photon's Electric and Magnetic fields and its Particle and Wave Characteristics. [Paper presentation]
'Annual Meeting of the Division of Nuclear Physics Volume 52, Number 10'. (2007, Oct 10-13). Moon, R., Vasiliev, V. Newport News, Virginia. USA.
American Physical Society; Abstract ID: BAPS.2007.DNP.BF.15
https://meetings.aps.org/Meeting/DNP07/Session/BF.15   http://meetings.aps.org/Meeting/DNP07
http://meetings.aps.org/link/BAPS.2007.DNP.BF.15

[35] The St. Petersburg State University experiment that discovered the Photon Acceleration Effect. 'Virtual Conference on Nanoscale Science and Technology' VC-NST. (2007, Oct 21-25). Moon, R., Vasiliev, V. University of Arkansas. 222 Physics Building. Fayetteville, AR 72701 USA.   http://www.ibiblio.org/oahost/nst/index.html

[36] The Explanation of Quantum Teleportation and Entanglement Swapping. [Paper presentation]
'49th Annual Meeting of the Division of Plasma Physics, Volume 52, Number 11' (2007, Nov 12–16). Moon, R., Vasiliev, V. Orlando, Florida. American Physical Society;
Abstract ID: BAPS.2007.DPP.UP8.21
https://meetings.aps.org/Meeting/DPP07/Content/901
http://meetings.aps.org/link/BAPS.2007.DPP.UP8.21
https://meetings.aps.org/Meeting/DPP07/Session/UP8.21

[37] The Explanation of the Photon's electric and magnetic fields, and its particle and wave characteristics. [Paper presentation]
'60th Annual Meeting of the Divison of Fluid Dynamics. Volume 52, Number 1' (2007, Nov 18–20). Moon, R., Vasiliev, V. Salt Lake City, Utah. American Physical Society;
Abstract ID: BAPS.2007.DFD.JU.22
http://meetings.aps.org/Meeting/DFD07
https://meetings.aps.org/Meeting/DFD07/Session/JU.22
http://meetings.aps.org/link/BAPS.2007.DFD.JU.22

[38] The Explanation of quantum entanglement and entanglement swapping. [Poster session]
'The 10th International Symposium on the Origin of Matter and the Evolution of the Galaxies (OMEG07)' (2007, Dec 4-6) Moon, R., Vasiliev, V. Hokkaido University, Sapporo, Japan.
Bibcode: 2008AIPC.1016.....S, Harvard (Astrophysics Data System) ISBN 0735405379
https://ui.adsabs.harvard.edu/abs/2008AIPC.1016.....S/abstract
http://nucl.sci.hokudai.ac.jp-omeg07/

Books by author {A} and work presented in other published books/booklets

1. *"The Vortex Theory of Matter"* Copyright 1990
   R. Moon. {A} Costa Mesa, California

2. *"The End of The Concept of Time"* Copyright 2000.
   R. Moon. {A} Gordon's Publications of Baton Rouge. Louisiana. ISBN 096792981-4.

3. *"The Bases of the Vortex Theory of Space"* (2002).
   R. Moon. {A} Publishing house; "ZNACK" Director Dr. I. S. Slutskin. Post Office Box 648. Moscow, 101000, Russia. p. 32. (In Russian). Journal ISSN: 2362945.

4. *"The Vortex Theory...The Beginning"* (2003). Copyright 2003.
   R. Moon. {A} (Editor's note by Prof., Dr. Victor V. Vasiliev)
   Gordon's Publications of Fort Lauderdale Fla. USA.

5. *"The Bases of the Vortex Theory"* (2003).
   Book of abstracts: Russian Academy of Sciences; ISBN 5-98340-004-5; TRN: RU0403918096768 OSTI ID: 20530263 R. p. 251. R. Moon. V. Vasiliev
   http://nuclpc1.phys.spbu.ru/nucl/Abstracts/Nucleus_2003.pdf
   http://physics.doi-vt1053.com/ISBN5-98340-004-5/Nucleus_2003.pdf

6. *"The Vortex Theory and some interactions in Nuclear Physics"* (2004).
   Book of abstracts: Russian Academy of Sciences; ISBN 5-9571-0075-7 p. 259.
   R. Moon. V. Vasiliev
   http://nuclpc1.phys.spbu.ru/nucl/Abstracts/Nucleus_2004.pdf
   http://physics.doi-vt1053.com/ISBN5-9571-0075-7/Nucleus_2004.pdf

7. The Vortex Theory Explains the Quark Theory. (2005).
   R. Moon. {A} Gordon's Publications of Fort Lauderdale, Florida. USA. p. 205.

8. Dr. Russell Moon PhD Thesis; *"The End of "Time"* Collection of Learned Works Addendum, 2012, (pp. 473-488) VVM Publishing House: ISBN 978-5-9651-0804-6 Editor in Chief: I. S. Ivlev. Saint Petersburg State University. St Petersburg, Russia.
   http://physics.doi-vt1053.com/ISBN978-5-9651-0804-6/Dr-Russell-G-Moon-PhD-thesis-The-End-of-Time.pdf
   http://physics.doi-vt1053.com/ISBN978-5-9651-0804-6/Natural_Anthropogenic_Aerosoles_4pages.pdf

9. *"The Discovery of the Fifth Force in Nature: The Anti-Gravity Force"* Collection of Learned Works (pp. 489-495) R. Moon. M. F. Calvo.
   VVM Publishing House: ISBN 978-5-9651-0804-6 p. 534. Editor in Chief: I. S. Ivlev. St Petersburg State University. St Petersburg, Russia.
   http://physics.doi-vt1053.com/ISBN978-5-9651-0804-6/The-Discovery-of-the-Fifth-Force-in-Nature:-The-Anti-gravity-Force.pdf
   http://physics.doi-vt1053.com/ISBN978-5-9651-0804-6/Natural_Anthropogenic_Aerosoles_4pages.pdf

10. *"The Discovery of the Fifth Force in Nature: The Anti-gravity Force"* Collection of Learned Works (pp. 496-503) V. Vasiliev. R. Moon. M. F. Calvo. VVM Publishing House; ISBN 978-5-9651-0804-6 2013. p 534. Editor-in-Chief: I. S. Ivlev, Saint Petersburg State University. St. Petersburg. Russia.
http://physics.doi-vt1053.com/ISBN978-5-9651-0804-6/The-Discovery-of-the-Fifth-Force-in-Nature:-The-Anti-gravity-Force.pdf

Other References

1) Stephen Smale. '*A Classification of Immersions of the Two Spheres*'. Transactions of the American Mathematical Society, Vol. 90, No. 2. (Feb 1959). Online; ISSN 1088-6850. Printed version; ISSN 0002-9947

2) H. Yukawa, "Tabibito" (The Traveler) World Scientific, Singapore. (1982) pp. 190-202. ISBN: 9971950103

3) Wolfgang Pauli. Nobel Lecture; for the discovery of the exclusion principle. Stockholm, Sweden, (1946).
https://www.nobelprize.org/uploads/2018/06/pauli-lecture.pdf

4) Robert Desbrandes, Daniel Van Gent. '*Intercontinental quantum liaisons between entangled electrons in ion traps of thermoluminescent crystals*', (2006, 11. 09) arXiv:quant-ph/0611109.
https://arxiv.org/ftp/quant-ph/papers/0611/0611109.pdf
https://doi.org/10.48550/arXiv.quant-ph/0611109

5) D. Bouwmeester. Jian-Wei Pan, K. Mattle, M. Eibl, H. Weinfurter, A. Zeilinger. '*Experimental Quantum Teleportation*' Nature 390, 6660, 575-579 (1997),
https://doi.org/10.48550/arXiv.1901.11004
https://arxiv.org/pdf/1901.11004.pdf

6) Einstein, A. Podolsky, B. Rosen, N. '*Can Quantum-Mechanical Description of Physical Reality Be Considered Complete?*' Phys. Rev. 47, 777, (1935).
https://link.aps.org/doi/10.1103/PhysRev.47.777
https://journals.aps.org/pr/abstract/10.1103/PhysRev.47.777

7) David Halliday, Roberts Resnick. and Jearl Walker. '*Fundamentals of Physics*'; 7th Edition. (2005) ISBN 9780471216438
http://cdn.worldcolleges.info/sites/default/files/Instructors_Manual.pdf

8) Thomas S. Kuhn. '*The Structure of Scientific Revolutions*'. Chicago, Illinois: University of Chicago Press. p. 17. (1996) ISBN 10: 0226458083   ISBN 13: 9780226458083

9) Michael Marshall. '*Gigantic jets blast electricity into upper atmosphere*'. New Scientist, (August 23, 2009).
http://www.newscientist.com/article/dn17664-gigantic-jets-blast-electricity-into-upper-atmosphere.html

10) Mills, Allan. Part 6: The Leyden jar and other capacitors. Bulletin of the Scientific Instrument Society (UK) (99): (December 2008) pp. 20–22.
http://web.archive.org/web/20110727024546/
https://www.sis.org.uk/bulletin/99/mills.pdf

11) Benjamin Franklin, etal. Leonard. W. Labaree,ed. '*The Papers of Benjamin Franklin*' New Haven, Connecticut: Yale University Press, 1961) vol. 3, p. 352: Letter to Peter Collinson. (April 29, 1749). Paragraph 18.
http://franklinpapers.org/franklin/framedVolumes.jsp?vol=3&page=352a

12) H. E. Bond. E. P. Nelan. D. A. Vandenberg. G. H. Schaefer. D. Harmer. (2013)
Abstract: HD 140283: '*A Star in the Solar Neighborhood that Formed Shortly After the Big Bang*'. The Astrophysical Journal Letters. Volume 765 (1): L12. Bibcode:2013Apj…765L…12B.
https://iopscience.iop.org/article/10.1088/2041-8205/765/1/L12/pdf

Russian Scientific Journals:

1. http://www.new-philosophy.narod.ru/RGM-VVV-RU.htm (in Russian)
2. http://www.new-philosophy.narod.ru/MV-2003.htm (in English)
3. http://www.new-idea.narod.ru/ivte.htm (in English)
4. http://www.new-idea.narod.ru/ivtr.htm (in Russian)
5. http://www.new-philosophy.narod.ru/mm.htm (in Russian)

PART VI
THREE EXPERIMENTS PROVING THE VORTEX THEORY OF ATOMIC PARTICLES

The Ball and Zinc Experiment...

The Proof of "The Vortex Theory"
Konstantin Gridnev, Russell Moon, Victor Vasiliev

Institute of Physics, St. Petersburg State University Russia,
Independent Researcher, U.S.A., All Russian Electrotechnical Institute, RUSSIA.

ABSTRACT

According to the principles of the Vortex Theory, protons and electrons are three-dimensional holes connected by invisible fourth-dimensional vortices. It was further theorized that when photons are absorbed then readmitted by atoms, the photon is absorbed into the proton, moves through the fourth-dimensional vortex, then reemerges back into three-dimensional space through the electron. To prove this hypothesis, an experiment was conducted using a hollow aluminum sphere containing a powerful permanent magnet suspended directly above a zinc plate. Ultraviolet light was then shined upon the zinc. The zinc emits electrons via the photoelectric effect that are attracted to the surface of the aluminum sphere. The sphere was removed from above the zinc plate and repositioned above a sensitive infrared digital camera in another room. The ball and camera were placed within a darkened box inside a Faraday cage. Light was shined upon the zinc plate and the picture taken by the camera was observed. When the light was turned on above the zinc plate in one room, the camera recorded increased light coming from the surface of the sphere within the other room; when the light was turned off, the intensity of the infrared light coming from the surface of the sphere was suddenly diminished. Five other experiments were then performed to eliminate other possible explanations such as quantum-entangled electrons. These tests prove the existence of a new structure in nature: fourth-dimensional sub-atomic vortices. These vortices, named "Konsiliev Vortices" are proved by explaining previous unexplainable phenomenon of nature: the Conservation of charge, Entropy, the ½ spin of particles, Charge Conjunction, and Parity: Revealing a new level of scientific understanding above Relativity, and ushering in a new era in human history.

HISTORICAL BACKGROUND OF THE VORTEX THEORY:

The Vortex Theory is a new and revolutionary explanation for the construction of the universe. What follows is a brief synopsis of this theory.

Just as the Theory of Relativity is based upon the proposal that time is a fundamental principle of the universe and exists as a fourth-dimension called "space-time", the Vortex Theory is based upon the discovery that time is not a fundamental principle of the universe. Instead, time only exists as function of motion, a phenomenon created by motion.

The elimination of time as a fundamental principle of the universe reduces the five "pieces" of the universe to four: matter, space, energy, and the forces of nature. Because the construction of each "piece" of the universe depends upon the construction of the other pieces, when time is eliminated

- eliminating the fourth-dimension of "space-time" - not only does the construction of space have to be re-evaluated, but also, the other three: matter, energy, and the forces of nature.

In an effort to determine how space is constructed when "space-time" is eliminated, the premise that space is a void and the premise that space possesses a fourth-dimension was re-examined. During this investigation, it became apparent that if a pure fourth-dimension of space existed [a spatial dimension devoid of any "time" characteristics], one characteristic indicative of its presence would be the existence of three-dimensional holes. [Note: all holes are one dimension lower than the space they exist within.] While looking for these three-dimensional holes, it became intriguingly apparent that protons and electrons could indeed be these holes. When this hypothesis was used to explain the length shrinkage and time dilation effects associated with the Michelson Morley experiment, the mathematics proved it to be correct.

The mathematical analysis revealed that space was made of something and that matter exists as three-dimensional [3d] holes upon the surface of fourth-dimensional [4d] space. Even more startling, the mathematics further revealed that these holes in space are really the ends of vortices extending into and out of 4d space. The particles we call protons and electrons are 3d holes connected to each other via a 4d vortex. Three-dimensional space flows into the proton [creating its electrostatic charge], through the vortex, then exits at the electron [creating its opposite electrostatic charge].

When particles were originally created in the early universe, protons were attached to anti-protons via 4d vortices, while electrons were similarly attached to positrons. However, in the early universe, how these vortices then became cross-connected, broke, and reattached protons to electrons and anti-protons to positrons is beyond the scope of this paper. Suffice it to say that when 4d vortices cross, they break, and the ["particles"] holes become connected to those at the opposite ends of the other vortex. This break and reconnection appears to be responsible for the law of the Conservation of Charge.

There are at least six types of vortices:

(1) Artificially created unstable vortices existing between quantum entangled particles possessing opposite spin states, such as electron/electron pairs, proton/proton pairs, or photon/photon pairs;
(2) Original 4d vortices such as those existing between particles and their anti-particle (meson/anti-meson pairs, lepton/anti-lepton pairs, baron/anti-baron pairs);
(3) Higher dimensional vortices (dimensions 5-7) existing between quark/anti-quark pairs;
(4) "Three-dimensional" vortices existing from electrons to protons in atoms;
(5) The broken then reconnected "cross-connected" 4d vortices such as those existing between protons and electrons, and anti-protons and positrons;
(6) The vortices twisted into a 4d torus around particles such as neutrons [such twisted vortices also appear to exist around strange quarks].

However, only the broken and reconnected vortices existing between the protons and electrons of matter are the subject of this paper.

These broken, then, reconnected vortices are most prevalent within atoms and between the opposite charges of ions. Within atoms, protons and electrons are attached to each other by "cross-connected" vortices flowing from the proton to the electron in 4d space. The manner via which this cross connection occurs is as follows:

When a hydrogen atom is created, some of the 3d space flowing out of an electron begins to flow into a proton creating an electrostatic attraction via Coulomb's law. As these two holes move closer, more and more space flowing out of the electron flows into the proton until a critical distance is reached where all of the 3d space flowing out of the electron now flows directly into the proton. When this situation occurs, a 3d vortex in our 3d space is created. Then, as the electron begins to rotate about the proton, the two 4d vortices connecting them to other particles in the universe cross, break, and reconnect; causing the proton and the electron to be connected together. At the same instant, the other two broken ends connect to each other, joining the other two particles [holes] together. This sequence is seen in the following illustrations of Figures 1 – 5.

Figure 1

The electron and the proton are attracted to each other via the electrostatic charge. The thick dark lines represents the 4d vortices flowing from the proton to the anti-proton and from the positron to the electron.

Figure 2

When the electron and the proton are close enough, all of the space flowing out of the electron flows into the proton forming the 3d vortex [thick dotted line], and the electron begins to "revolve", move about the proton.

Figure 3

The proton and the electron begin to revolve around each other.

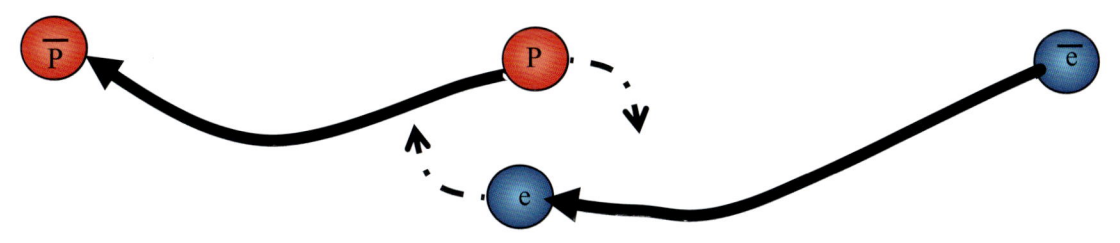

Figure 4

The 4d vortices cross and break.

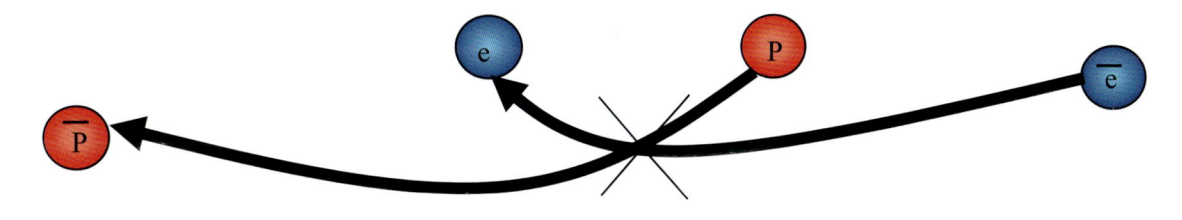

Figure 5

The electron is now connected to the proton and the positron is connected to the anti-proton.

The two vortices between the electron and the proton now create a circulating flow containing a fixed volume of space. This circulating volume of 3d space continually flows from the proton, into 4d space - through 4d space, and then into the electron. Here, it exits the electron, flowing back through 3d space and into the proton once again, binding the proton to the electron creating a hydrogen atom.

Figure 6

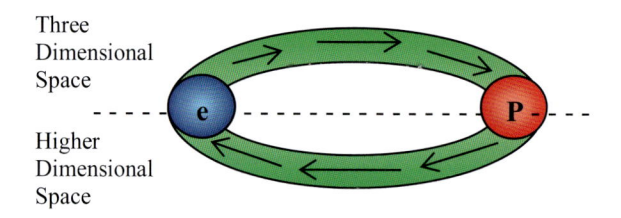

When the circulating flow commences, both of the electrostatic charges are neutralized. The word "neutralized" is used because no flowing space escapes from the system.

[It is necessary to mention that the neutron is a vortex caught in a loop. The neutron is theorized to be a proton completely surrounded by an electron: a hole within a hole. The neutron can also be created when the proton's vortex breaks, causing the end nearest it to surround it creating a neutron, while the other broken end emerges into 3d space as a positron. As such, the vortex surrounding the proton is turned into a 4d torus: a 4d vortex turned inside-out (similar to a smoke ring though impossible to draw).] Also, according to this model, a photon is a packet of condensed 3d space whose frequency of vibration is a function of the amount of condensed 3d space contained within it.

When a photon is absorbed then readmitted by an atom, it is theorized that the photon is absorbed by the proton; its volume suddenly adding to the length of the vortex. If the photon is the right frequency [possesses the right volume], the electron jumps into a higher "orbit" and stays there due to a balanced Coulomb force. If the photon does not possess the right volume, the Coulomb force is unbalanced and the electron is pulled backwards to a lower "orbit" as it readmits the photon back into 3d space.

When an atom captures or loses an electron and becomes an ion, the 3d vortex is broken and the electron is released from the atom. However, because both proton and electron are merely the two 3d ends of a 4d vortex, the 4d vortex between them remains. Consequently, a photon of energy entering into the proton will still travel through the 4d vortex to the electron at the other end no matter where the electron is located.

THE EXPERIMENT

If this theory is correct in its assumption that a photon is absorbed by the proton and then readmitted through the electron, this hypothesis can be verified by the use of the photoelectric effect.

Using the photoelectric effect, electrons can be discharged from a zinc plate using ultraviolet light. If a sphere is placed directly above the plate, the upward accelerated electrons will adhere to the outside surface of the sphere. Then, if light is shined upon the zinc, photons should travel into the protons, through the 4d vortex, and out of the electrons. If the electrons can be visually monitored, and if light is seen coming out of them, the existence of the vortices can be proven.

The experiment to prove the Vortex Theory is as follows. A zinc plate is placed just below an aluminum sphere, almost touching it. Ultraviolet light is then shined upon the point where they almost touch. Electrons via the photoelectric effect are accelerated off of the zinc plate and onto the sphere. The sphere is then carefully removed from its position above the plate, and taken to another darkened room where it is placed upon a stand directly above a sensitive infrared camera. The stand, ball, and camera are then placed inside a darkened cardboard box that is located inside of a Faraday cage. Then, a light bulb is alternately turned on and off above the zinc plate. If the infrared camera photographs an increase in light when the bulb is on and a decrease when the light is turned off, the vortices exist, and the Vortex Theory is true.

SETTING UP THE EXPERIMENT

To prove the existence of the vortices and with them the Vortex Theory, the following experiment was devised for both its simplicity and cost effectiveness:

A thin walled 19 cm diameter aluminum sphere has a 5 cm hole cut in the top. Within the hole is placed a powerful 0.3 Tesla permanent magnet with the south end facing the zinc plate. The sphere is then suspended using a 10 cm thin piece of wood tied to a wire, placed inside the hole, and attached to a 45 cm long top wood support. The sphere needs to be suspended a distance of .3mm above a 15 cm by 31 cm plate of zinc, previously sanded: the zinc plate is sanded using fine sandpaper and wiped clean to remove the top oxidized layer; and then wiped clean using a damp cloth. [Although the other wood supports are permanently attached to each other, the "Top wood support" merely sits upon the two vertical supports allowing it to be easily removed and replaced.]

Figure 7

Top view

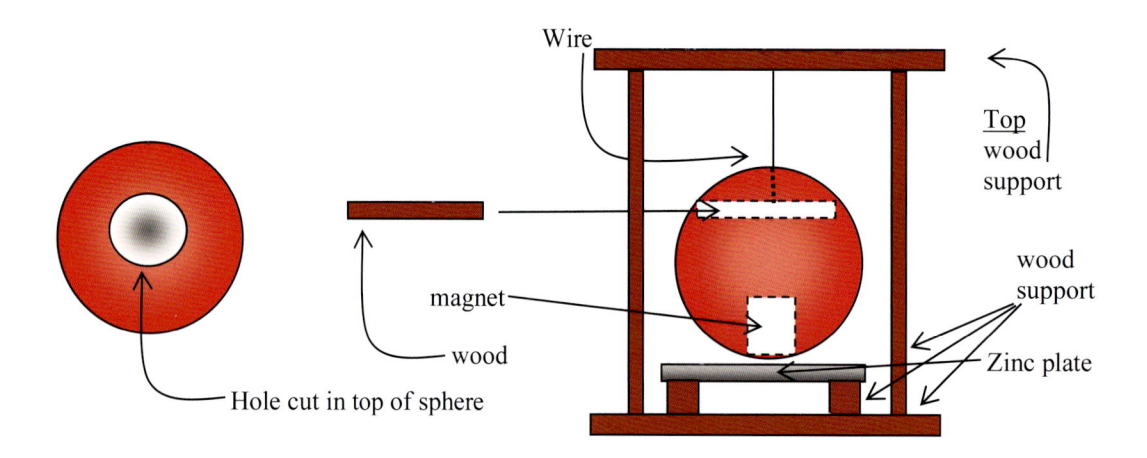

A second apparatus seen in Figure 8 is constructed. The camera is an Orion StarShoot Deep Space Color Imaging Camera #52065 with a Celestron 25mm SMA 1.25 inch Wide Angle Lens Model number 93007-A attached to it [the infrared filter is removed to make it more sensitive]. The apparatus is then placed inside a cardboard box sized: 45 cm by 45 cm, by 32 cm.

Figure 8

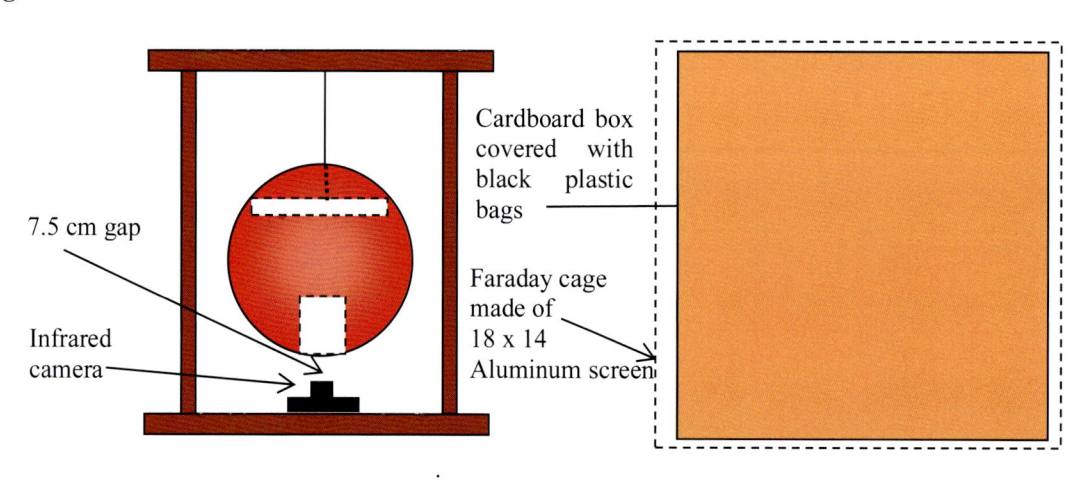

The cardboard box is covered by two layers of 3mil thick black plastic to keep light out of the inside of the box. A small hole for the 124 cm electronic cable that attaches the infrared camera to the laptop computer through the USB port is cut in both the box, plastic, and Faraday cage [made of 18 x 14 aluminum screen]. The tiny space between the wire and cardboard is covered with additional screen and taped in place with duct tape. A ground wire is attached from the screen to the ground [earth] connection on an electrical wall plug or to a grounded metal water pipe.

The software for the Orion StarShoot camera is installed on the computer. After installation and after the program is on the viewing screen, the CAMERA CONTROL window settings are placed at: MODE – [light mono 2 x 2]; SECONDS [60]; DARK SUBTRACT [off]; SETUP [Analog gain

63, Analog offset 100]. The SCREEN STRETCH window is set to [medium]; the TEC thermoelectric cooler is off.

THE EXPERIMENT

Three lamps with 75-watt ultraviolet light bulbs have their lamp shades removed, and are then positioned so the light bulbs are 15 cm from the tangent point where the sphere almost touches the zinc plate and are left on for 20 minutes.

Afterwards, the lamps are turned off and the room is darkened, the top wooden support is carefully lifted up, lifting the aluminum sphere with it. Being careful not to let the sphere touch any object, it is carried to another room and placed upon the supports of the second apparatus within the box.

The lids of the box are folded shut and the ends of the black plastic are folded over the top of the box to seal it [a large book can be placed on top of the box to the keep the box closed].

Next, with the lights turned out in the room containing the zinc plate, the time is recorded in a ledger, the "Expose" button on the camera control window is clicked with the left side of the mouse, and the exposure begins. After 60 seconds the exposure is ended; the picture pops up on the computer screen, and the display on the screen stretch window appears. The minimum and maximum values are recorded in the ledger, and the picture is assigned a designation and saved.

The lights are then turned on above the zinc plate in the other room [3 – 75W uv bulbs, 1 - 75W "Blue" daylight bulb, and 1 – 75W compact florescent daylight bulb]. The "Expose" button on the camera control window is again clicked and the procedure is repeated. It is assumed that the time between tests takes approximately two minutes.

This procedure is repeated at least ten times. A typical result is as follows:

Table 1

August 5, 2009 [TEC camera cooling device turned off] Temp. 79^0 F, 41% Humidity.

Lights	Time pm	Min	Max	Lights	Time pm	Min	Max
#1 Off	10:59	12397	15084	#11 Off	11:18	11843	13843
#2 On	11:01	11862	13871	#12 On	11:20	11911	13979
#3 Off	11:02	11806	13779	#13 Off	11:21	11885	13919
#4 On	11:05	11864	13882	#14 On	11:23	11920	13999
#5 Off	11:06	11813	13782	#15 Off	11:25	11873	13898
#6 On	11:08	11881	13915	#16 On	11:27	11914	13984
#7 Off	11:10	11835	13828	#17 Off	11:29	11876	13906
#8 On	11:12	11888	13932	#18 On	11:31	11939	14040
#9 Off	11:14	11875	13898	#19 Off	11:33	11903	13960
#10 On	11:16	11911	19978				

ANALYSIS:

Note how test #1 was higher than test #2 even though the lights were off in test #1. This result was determined to be an effect created by the camera. [When the ball was removed from the box and after waiting for 30 minutes, the first picture has a high pixel reading.]

Also note how the min. and max. increase when the lights are on and decrease when the lights were off. This particular test was repeated 26 times. The first 18 tests were with the lights alternately off and on. In tests #19-26 the lights were left off. When off, the previous alternately rise and fall of the min. and max. values was suddenly destroyed: see Table #1 continued:

Table 1 continued

Lights	Time pm	Min	Max	Lights	Time pm	Min	Max
#20 Off	11:35	11904	13965	#24 Off	11:43	11928	14017
#21 Off	11:37	11981	14136	#25 Off	11:46	11884	13931
#22 Off	11:39	11937	14044	#26 Off	11:47	11859	13885
#23 Off	11:41	11890	13940				

ANALYSIS:

Note the different results in Table #2 from the first 19 tests seen in Table #1. Although in test #20, the min. and max. are higher than those in #19, the rise is only one pixel. Next, test #21 is now higher than test #20. If the same sequence was being followed here that occurred in the first 19 tests, #21 should be lower, and test #22 that should be higher than #21 is instead lower. Consequently, after the lights were shut off, instead of seeing the usual higher-lower sequence, we see after #19 the sequence is disrupted: #20 [just slightly] higher, #21 higher, #22 lower, #23 lower, #24 higher, #25 lower, #26 lower.

Apparently, turning off the lights in the other room for the last eight tests disrupted the sequence. *It is important to note, that according to conventional physics, no such sequence should have ever existed in the first place.*

A second way to analyze the data is to look at the total intensity of the pixels.

In this version using the program called Maxim, the brightness of every pixel in the image is measured. Then the values of all the pixels are added up. Because the imaging chip in this camera is 376 by 291 pixels, there are 109,416 pixels. Example: on image 'D', the total intensity = 1.316829E9 (or 1,316,829,000) With 109,416 pixels, that means an average brightness of 12035.066 per pixel. Using this system we are able to compare the brightness of the pixels from when the camera was either turned on or off.

Table 2

A	7/17/2009	2:59:19	1.380234E+09	N	7/17/2009	3:23:59	1.323555E+09
B	7/17/2009	3:01:33	1.316550E+09	O	7/17/2009	3:25:28	1.317972E+09
C	7/17/2009	3:02:59	1.310113E+09	P	7/17/2009	3:27:30	1.322917E+09
D	7/17/2009	3:05:10	1.316829E+09	Q	7/17/2009	3:29:02	1.318437E+09
E	7/17/2009	3:06:35	1.310952E+09	R	7/17/2009	3:31:25	1.325845E+09
F	7/17/2009	3:08:52	1.318979E+09	S	7/17/2009	3:33:12	1.321489E+09
G	7/17/2009	3:10:24	1.313633E+09	T	7/17/2009	3:35:03	1.321712E+09
H	7/17/2009	3:12:33	1.319806E+09	U	7/17/2009	3:37:58	1.331010E+09
I	7/17/2009	3:14:19	1.318168E+09	V	7/17/2009	3:40:17	1.325814E+09
J	7/17/2009	3:16:34	1.322614E+09	W	7/17/2009	3:41:52	1.319998E+09
K	7/17/2009	3:17:55	1.314442E+09	X	7/17/2009	3:43:56	1.324585E+09
L	7/17/2009	3:20:05	1.322519E+09	Y	7/17/2009	3:45:27	1.319401E+09

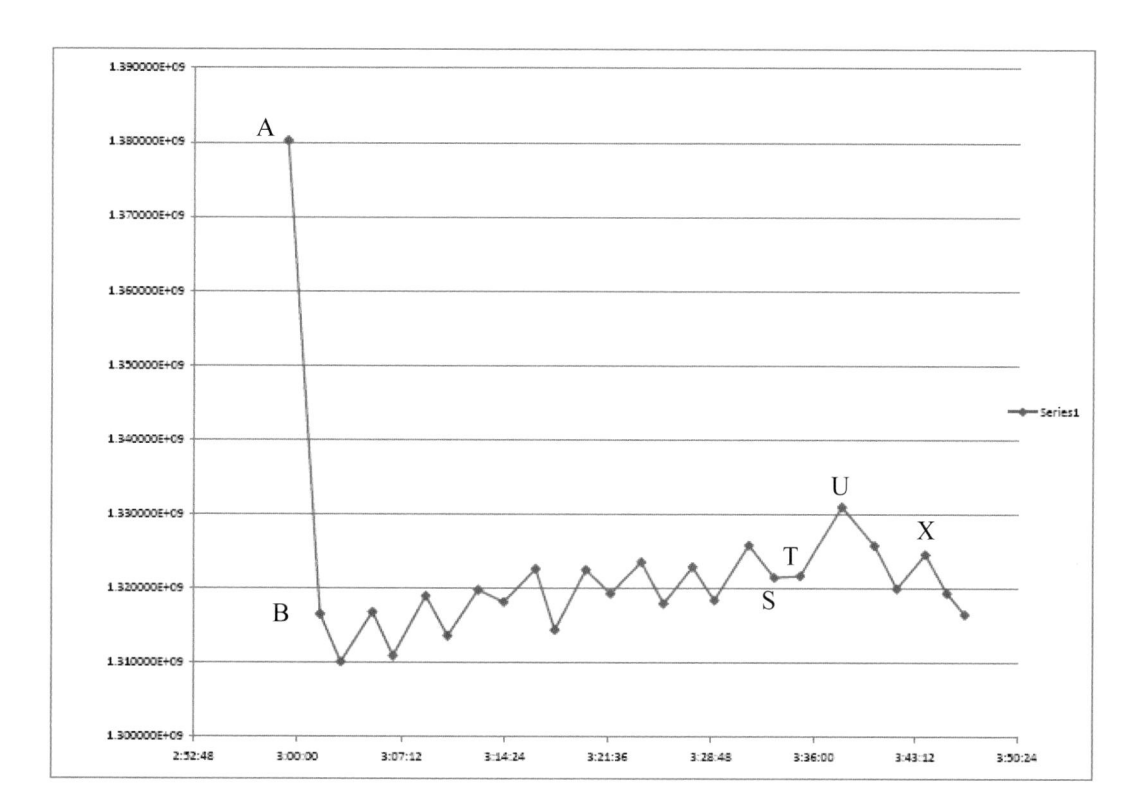

Figure 9 The data from the pixel brightness is graphed below.

This graph reveals some fascinating insights into the dynamics of the vortices:

Notice at the start of the sequence at point **A**, even though the lights are off the reading was exceedingly high. Again, this is a phenomenon being created by the camera used in this experiment.

Notice too, that when the lights are turned on at point **B**, the rise and fall of the graph takes place along a narrow band of values in relation to the starting value. This reveals that a fairly even amount of light is being absorbed by the protons and re-admitted by the electrons. It is assumed but not known at this time that the steady increase in the graph, may be due to the heating of the camera in the box. When the lights are turned off at point **S**, but not turned back on at point **T** and left off for the rest of the experiment, notice how the steady sequence of increase and decrease is suddenly disrupted. This reveals that the turning off, of the lights in the other room was responsible for this change. The sudden increase in the values at **U** and **X** could be the result of photons - as noise from the zinc plate - being built-up in the vortices, and suddenly discharging through the electrons on the sphere, but this is conjecture.

ALTERNATE EXPERIMENTS A-E

In order to determine if an unforeseen variable was affecting the results of the tests and creating erroneous data, experiments were conducted to determine if the current in the wires supplying power to the camera and, or, the computer was creating the effect; or, if the camera itself was defective.

ALTERNATE EXPERIMENT A:

In order to determine if the current flowing through the wires in the building affected the computer or the camera [and hence the data] when the lights were being turned off and on, it was decided to make a series of tests with the ball removed from inside the box. [Note: if the ball was still present and noise from the zinc in the form of noise was still being transferred through it, it would affect the data.] The typical results were as follows:

Table 3

From August 1, 2009 [TEC camera cooling device turned off] 78^0 F, 42%Humidity

Lights	Time pm	Min	Max	Lights	Time pm	Min	Max
#1 Off	11:30	11521	13116	#6 On	11:40	11656	13429
#2 On	11:32	11571	13234	#7 Off	11:42	11672	13466
#3 Off	11:34	11598	13295	#8 On	11:44	11695	13514
#4 On	11:36	11616	13345	#9 Off	11:46	11707	13544
#5 Off	11:38	11638	13388	#10 On	11:49	11727	13583

ANALYSIS:

Notice how the values of the min. and max. steadily climbed up. [It is assumed because the camera-cooling device was turned off, that the values slowly rose as the camera slightly heated up due to its being inside of the sealed box.] Also notice how the alternate turning on and off, of the lights in the other room had no effect on the results. The values of the min. and max. did not alternately rise and fall as they did in Table #1.

ALTERNATE EXPERIMENT B:

In order to determine if the camera was defective and was causing the results in Table #1, it was decided to test the camera only. Because the values in Alternate Experiment A continually rose with the TEC camera cooling device turned off, it was decided to make another series of tests with the TEC cooling device turned on. It was predicted that the values should steadily fall as the camera slowly cooled.

Again, with the ball removed from the box, the typical results were as follows:

Table 4

From August 3, 2009 [TEC camera cooling device turned on] 78 F, 42% Humidity

Lights	Time pm	Min	Max	Lights	Time pm	Min	Max
#1 Off	9:55	10692	11154	#6 Off	10:00	10529	11057
#2 Off	9:56	10650	11128	#7 Off	10:02	10511	11050
#3 Off	9:57	11613	11103	#8 Off	10:03	10488	11042
#4 Off	9:58	11581	11097	#9 Off	10:04	10468	11074 ?
#5 Off	9:59	10557	11072	#10 Off	10:05	10450	11027

ANALYSIS:

Notice the difference from Table #3. Here, instead of continually rising, the values steadily fall as the camera slowly cools off. The only odd result is the increase in the max. value of test #9.

ALTERNATE EXPERIMENT C:

Because the values of the min and max continued to rise in Alternate Experiment B with the lights kept off and TEC camera cooling device turned on, it was decided to run another series of tests with the TEC off and the lights off. It was predicted that the values of the min and max would steadily rise as the camera heated up.

Table 5

From August 3, 2009 [TEC camera cooling device turned off] Temp 78^0 Humidity 41%

Lights	Time pm	Min	Max	Lights	Time pm	Min	Max
#1 Off	10:55	11445	12938	#6 Off	11:02	11489	13082
#2 Off	10:56	11458	12983	#7 Off	11:02	11505	13114
#3 Off	10:58	11457	13002	#8 Off	11:03	11513	13134
#4 Off	10:59	11464	13024	#9 Off	11:04	11521	13155
#5 Off	11:00	11483	13062	#10 Off	11:05	11532	13174

ANALYSIS:

The camera acted as predicted. The values continued to rise as the test progressed.

ALTERNATE EXPERIMENT D:

The purpose of Alternate Experiment D was to use light of different frequencies to determine if all or only some make it through the vortices. Because the effect continued to work with infra-red, ultra-violet, and the light coming out of a "day light" bulb [a bulb that puts out all frequencies of light and is used for growing plants indoors] it is apparent that almost all of the light in the visual spectrum passes through the vortices. However, exactly how much and what is the most efficient range of frequencies will have to be determined by future experiments.

ALTERNATE EXPERIMENT E:

In order to eliminate the possibility that the effect we saw in the original experiment was being created by quantum entangled electrons and not the vortices between protons and electrons, this last experiment was conducted with the *roles of the ball and the plate reversed*. Instead of the light being shined upon the zinc plate and the camera taking a picture of the sphere, the camera was positioned above the zinc plate, and the light was shined upon the aluminum sphere.

It was necessary to conduct this experiment to eliminate the possibility that electrons being released from the zinc plate and transferred to the sphere were quantum entangled with other electrons on the plate. According to the principles of the Vortex Theory, an artificially created vortex exists between two quantum entangled electrons of opposite spins. When the electrons are separated from each other, the artificially created vortex continues to exist, connecting them to each other. In this situation, a photon of light hitting one electron excites this electron causing motion that is transferred down the vortex to the other electron causing a similar motion, causing it to release photons.

If this is what is happening in the original experiment, then it does not matter which end of the vortex the camera is looking at. Light shined on the sphere will excite the electrons on the sphere; this motion will be transferred down the vortex and back to the electrons at the other end on the zinc plate, exciting them, causing them to release infrared photons the camera can photograph: photons that would appear to be emanating from the zinc plate.

Or, just the opposite can occur. Light shined on the zinc plate will excite the electrons on the zinc, causing these motions to be transferred down the vortices to the electrons on the aluminum sphere, exciting them, causing them to release photons that the camera can photograph: creating the effects we have seen in our previous experiments.

To eliminate or confirm this possibility, the roles of the zinc plate and the aluminum sphere are simply reversed. The experiment starts out just as it did originally. The ball is positioned above the zinc plate. The UV lights are shined upon the position where the ball almost touches the plate. Then the Top wood support with the ball attached to it is carefully removed and the sphere is positioned between two raised supports [can be the backs of chairs]. The UV lights are positioned directly under the sphere and to the sides, as are the two daylight bulbs.

The entire apparatus containing the zinc plate is then carefully lifted up by the wooden supports and taken into another room. The camera is then placed upon other supports so that it can be positioned directly above the zinc plate. The camera is now focused upon the exact point where the sphere was closest to it, and at the same distance it was from the sphere [7.5 cm]. The lights in the room containing the camera are turned off. Then, in the room containing the sphere, the lights are alternately turned off and on as before and the results are tabulated.

If the original experiment was correct, [if light is being transferred from the protons on the zinc to the electrons on the sphere via 4d space], then no effect should be seen when the lights are turned off and on in the other room containing the sphere. However, if we see the same effect as in the original experiment, if the min and the max values increase and decrease as the lights are alternately turned off and on, the conclusions of the original experiment are false. Also, because the camera is now outside of the cardboard box, it was assumed at the start of these tests that the results will be lower each time because the heat from the camera is being allowed to dissipate into the much larger volume of air in the room. The results were spectacular:

Table 6

From August 12, 2009 [TEC camera cooling device off] Temp. 79^0, 41% Humidity

Lights	Time pm	Min	Max	Lights	Time pm	Min	Max
#1 Off	9:08	12147	14431	#8 On	9:27	11697	13437
#2 On	9:12	11790	13632	#9 Off	9:29	11680	13407
#3 Off	9:15	11733	13516	#10 On	9:32	11681	13405
#4 On	9:17	11732	13507	#11 On	9:33	11658	13354
#5 Off	9:20	11712	13475	#12 On	9:35	11638	13317
#6 On	9:22	11710	13475	#13 On	9:37	11628	13300
#7 Off	9:24	11701	13447				

ANALYSIS:

Just as before, the first test #1 with the lights off was high. [Again this effect is created by the camera being used in this experiment.] But now, notice that unlike the original test where the Min and Max values rose and fell when the lights were alternately turned off and on, in this test, there is a steady decrease in the light from test #2 to test #10. Notice too, how the lights from test #10 to test #15 were left on. This was done to see if the steady decrease in the values of the tests would be halted and an increase would occur when the lights were constantly left on, heating up the sphere. But, as can be seen, no change in direction occurred.

It must also be mentioned that there is a possibility that the results of all of these tests involved a combination of light traveling through the vortices from the protons to the electrons and light traveling through the vortices from quantum-entangled electrons. However, such future tests must be left to future experimenters, using much more technically advanced equipment able to distinguish between the two different types of vortices.

DISCUSSION:

The reports from the experiments performed have revealed two of nature's greatest secrets: that protons and electrons are holes in space; and the existence of tiny vortices of space flowing between subatomic particles. The proof of this statement is found in the six experiments performed in this paper and in the explanations of: <u>the ½ intrinsic spin of particles; the explanation of Charge conjunction; the explanation of the Conservation of Charge; the explanation of Entropy, and the explanation of Parity.</u>

The original experiment revealed that photons of energy were being transferred from the protons on the zinc plate to the electrons on the aluminum sphere. This phenomenon could only have happened if flowing vortices of 3d space exist between the proton and the electron in 4d space.

[Note, 3d space can flow through 4d space. For example, just like the relationship between a 3d vortex and its 2d interior surface, the interior of a 4d vortex possesses a 3d surface. The 3d volume flows into the 3d hole and becomes the surface of the 4d vortex. It flows along the length of the vortex, and when the end of the 4d vortex reemerges back into 3d space, it is the 3d surface of the 4d vortex that flows back into the 3d volume.]

If this phenomenon were occurring in 3d space, if say the effect was some type of electromagnetic phenomenon, it would be eliminated by the presence of the Faraday cage. The Faraday cage intercepts electromagnetic waves and shunts them to ground. Hence, the Faraday cage eliminates an explanation for the effects of the experiment being caused by some sort of electromagnetic wave.

Also, because both the zinc plate and the aluminum sphere were located upon wooden stands sitting upon floors covered with thick carpets, the argument that somehow minute vibrations being generated by the motions of atoms in the zinc are being transmitted through the floor to the electrons on the sphere is eliminated. If such a situation occurred, transmitting and receiving antennas would not be needed in modern communications.

The air is also eliminated. If somehow, in some unforeseen way, subatomic vibrations of the atoms on the zinc were affecting the molecules of the air, and these vibrations were traveling [much like sound] through the atmosphere to the electrons on the sphere, such an effect would be present in all of today's electronic circuitry. Its disruptive effect would be astronomical during daylight hours. The electronic circuitry of television and radio receivers would have to be placed within vacuum containers. Hence, this bizarre explanation can be eliminated because we would have already have witnessed its effects.

In alternate experiments A – D, other possible explanations were eliminated. Again experiment "A" revealed that the turning on and off of the lights did not create some sort of current change in the wires of the building that in turn affected the operation of the camera or the computer. Alternate experiment "B" eliminated the possible explanation that some undetected defect in the camera was causing the effect; and in conjunction with experiment "C", the possibility that the TEC cooling device being turned on or off was eliminated. Alternative experiment "D" revealed that apparently, almost any frequency of visual light is able to be transmitted. And finally, and most important of all, experiment "E" revealed that the effect was not being created by quantum entangled electrons.

In experiment "E", we saw that the camera did not photograph an increase in the radiation coming out of the zinc plate when the aluminum sphere in the other room had the lights turned on it. The light seen by the infrared camera in this experiment did not alternately rise and fall as it did in the original experiment when the lights were alternately turned on and off above the zinc plate, causing an increase in the light coming out of the aluminum sphere. Instead, this experiment confirmed the hypothesis that the photon cannot travel from the electron to the proton in 4d space: indicating that we are looking at a "one way trip",

Consequently, because of the above experiments, and since the ground, air, and electromagnetic emanations are eliminated, we have eliminated three-dimensional causes. Using the physics of the last century, we are left with no explanation whatsoever. However, using the principles of the Vortex Theory we have an explanation. We start with holes in space:

THREE DIMENSIONAL HOLES IN SPACE:

The mathematics of the Vortex Theory [Mr. Moon's PHD thesis] reveals that protons and electrons have to exist as holes in 3d space connected by 4d vortices. This mathematical analysis not only gives an exact explanation for the length shrinkage and time dilation effects of the Michelson Morley experiment, but also, an exact explanation for how the phenomenon of time is created and the explanation for its curious relationship with the speed of light.

The idea that protons and electrons are holes in space is not conjecture. The discovery that protons and electrons – and all particles possessing ½ intrinsic spin [except the neutrino] – are holes in space was already proven in the "Stern Gerlach" experiment. Unfortunately, the true significance of this great experiment was overlooked because the prevailing preference of thought then, [and today], was the idea that protons and electrons were "particles". To understand why "particles" possessing ½ spin are really holes in space, we must reexamine the Stern Gerlach experiment:

In 1921, two scientists, Otto Stern and Walter Gerlach performed an experiment that showed the "quantized" electron spin occurred in only two orientations.

The experiment was made using a beam of silver atoms that were heated in an oven, then shot through a non-uniform magnetic field, and impacted upon a photographic plate. Because the outer electron of the silver atom is in effect shielded by the other 46 electrons, the silver atom allowed them to study the magnetic properties of a single electron. However, because the outer electron has a zero orbital angular momentum, no such interactions between the silver atoms and the magnetic field were expected. What Stern and Gerlach expected to see was a "continuous smear" on the photographic plate. But this was not to be.

When the silver atoms were directed through the non-uniform magnetic field, instead of seeing a "continuous smear" upon the plate, Stern and Gerlach saw that the beam was split into two separate parts. This split indicated that there were just two possible orientations for the magnetic moment of the electron.

This result seemed impossible, because how can the outer electron of the silver atom obtain a magnetic moment if it has no angular momentum and hence cannot form a current loop that creates a magnetic moment?

The problem was (supposedly) solved four years later when Samuel Goudsmit and George Uhlenbeck proposed that the electron must possess *an intrinsic angular momentum*. Today this intrinsic angular momentum is called electron spin: (or ½ spin).

But instead of solving the problem, ½ spin only deepens the problem. Because how *can* the electron only spin in one of two possible orientations? (Called spin "up" and spin "down".) Furthermore, how can an electron, or for that matter a proton, neutron, neutrino or any other 'particle" that also possesses ½, spin, spin in only one of two possible orientations? It doesn't seem to make any sense! It is still one of the greatest and most perplexing mysteries in all of today's science! But not any longer!

According to the Vortex Theory, protons, electrons, and neutrons (along with a great number of other "particles") are not particles at all. Instead, they are three-dimensional holes existing upon the surface of fourth-dimensional space. Because each higher dimension is at right angles to all lower dimensions, *fourth-dimensional space is not only at right angles to three-dimensional space, it is at right angles to all three lower dimensions simultaneously.*

Because fourth-dimensional [4d] space is at right angles to three-dimensional [3d] space, *a 3d hole that allows entrance into higher dimensional space is at right angles to 4d space.* Therefore, its 3d surface forms a ninety-degree angle to a 4d axis of rotation that extends into 4d space.

Because a 4d axis of rotation is impossible to draw, a 2d to 3d representation is shown below:

Figure 10

Three-dimensional view:

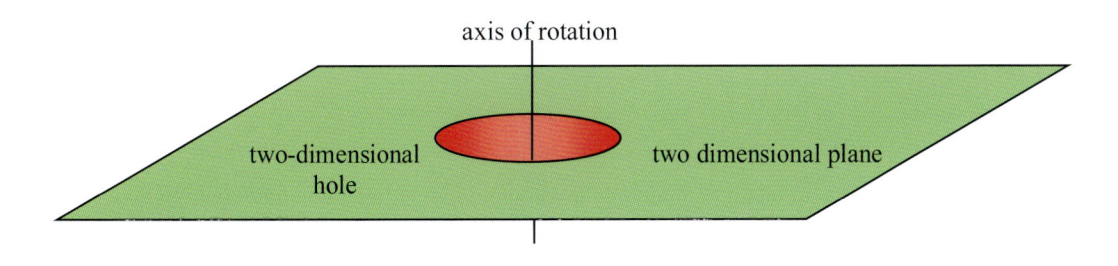

Looking at Figure 10 it becomes obvious that along the axis of rotation, there can only be two possible rotations: clockwise and counterclockwise, as seen below in Figure 11.

Figure 11

Top view:

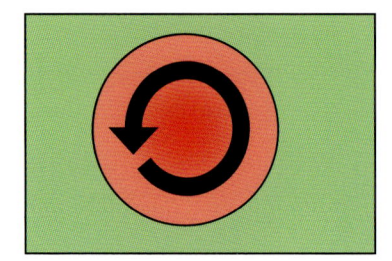

From this view you can clearly see why there are only clockwise or counterclockwise rotations.

In Figure 10, note how the 2d hole's 3d axis of rotation, that extends through its center and into 3d space, is at right angles to the surface of the 2d hole. Because of this right angle orientation, the hole can only rotate [spin] in one of two possible ways, clockwise or counter-clockwise. Even though the above illustration is a 2d to 3d illustration, the exact same situation – though impossible to draw – happens between 3d and 4d space.

Because a 3d hole possesses a 4d axis of rotation, it can only rotate clockwise or counter clockwise around its *fourth-dimensional axis*. Hence, it can only rotate in one of two possible spin orientations [that observers have designated "up" and "down"].

FOURTH DIMENSIONAL VORTICES:

Like the holes in space, the existence of the vortices were revealed by the mathematics of the Vortex Theory. This mathematics reveals that protons and electrons are connected by 4d vortices. But there is observational evidence too. This observation is found in the explanation of the Conservation of charge, the Charges of "particles", Entropy, and Parity.

PARITY

According to the model of the universe proposed by the Vortex Theory, the intrinsic spin of "particles" [holes in space] is a function of the direction of their orientation with regards to 4d space. Because it is impossible to draw 4d space, the 2d to 3d relationship is again drawn:

Figure 12

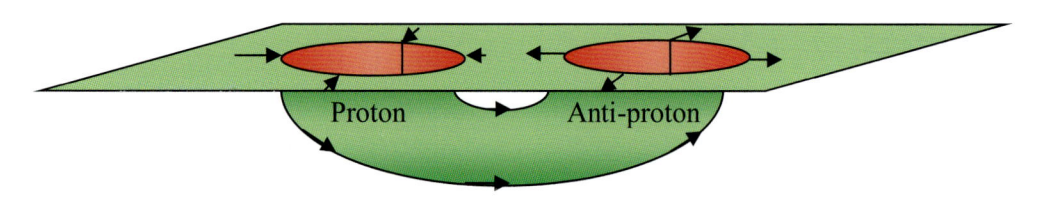

In the above drawing, note how the 3d space is flowing into the proton and out of the anti-proton. Note too how the space is flowing in opposite directions to each other – creating the opposite effects.

Again, applying the above relationship to 3d and 4d space, it can be seen that when the space within the vortex is rotating, opposite spin states are created at the opposite ends of the vortex. Because the vortex ends are connected to the same surface, a clockwise rotation in one end produces a counter-clockwise rotation in the other end. For example:

Figure 13 [Top view of Figure 12]

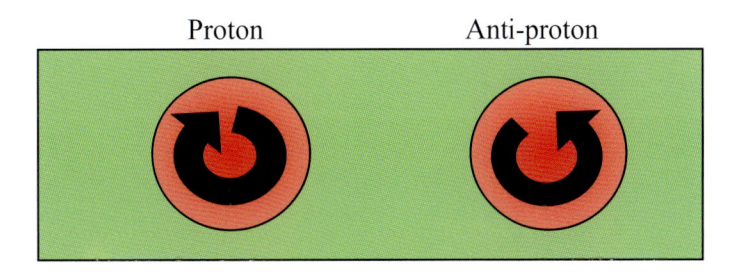

In the above drawing it can now be seen how the rotation of space within the vortex creates the opposite spin states of a particle and its anti-particle. Also because the vortex model proposes that magnetism is nothing more than the rotation of space, it can now be seen how the opposite magnetic moments of the two "particles" are also created".

CHARGE CONJUNCTION

According to the model of the universe proposed by the Vortex Theory, the charge a particle possesses is created by the space flowing into or out of a particle. For example, in the proton anti-proton pair, 3d space flows into the 3d hole of the proton then outwards into 4d space, through 4d space, into the anti-proton where it again exits into 3d space. This relationship can be seen in the 2d to 3d relationship drawn below:

Figure 14

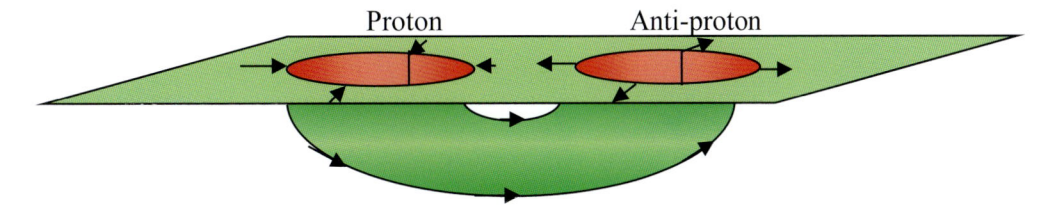

In the above drawing, notice how the two dimensional [2d] space flows into the proton through the 3d vortex then back through the anti-proton and out onto the 2d plane. Note too, how the 2d space is really the surface of a volume of 3d space contained within the vortex.

Applying the above relationship to 3d and 4d space, if the 3d space that flows into one "particle" flows out of the other "particle", it can be seen that the Charge Conjunction is a result of the 3d space flowing between the two particles in the form of a vortex.

THE CHARGES OF "PARTICLES":

It is an age-old observation that the charges of protons and electrons are opposite to each other. Although science presently has the charge pointing out of the proton and into the electron, this is a mistake. These directions were originally assigned over 200 years ago by Benjamin Franklin.

According to the principles of the Vortex Theory in its explanation of the Conservation of Lepton Number, space is flowing into the electron and out of the positron. And according to the Vortex Theory's explanation of gravity, space is flowing into the proton, pulling space into it; surrounding it with a massive region of less dense space [neutrons too]. When added together and viewed from our perspective here on earth, they create a seemingly massive region of less dense space. This region of less dense space is seen as Einstein's "bent" region of space surrounding planets and stars. As a consequence of these observations, it is apparent that space appears to be flowing into the proton.

Because of their opposite charges, if space is flowing into the proton then it is flowing out of the electron. Since the proton and the electron are 3d holes in space, the space flowing into the proton is going into 4d space and the space coming out of the electron is coming out of 4d space. And just as is seen in Figure 14, the interior of a 2d hole encircles a 3d volume; while a 3d hole encircles a 4d volume connected by a 4d tubular structure. This structure is the vortex.

It is also natural to conclude that if the vortices should break in 4d space that the broken ends could reappear back onto the 3d surface with space flowing out of the end connected to the proton and space flowing into the end still connected to the electron. This hypothesis is confirmed in the observations that revealed the existence of a law of nature designated by physicists as "The Conservation of Charge".

THE CONSERVATION OF CHARGE:

The term "conservation of charge" is based upon the scientific observation that the break-up of a "particle", or the change in a particle from one type to another [such as the decay of the neutron into a proton, electron, and an anti-neutrino] always leaves the same net charge. For example, since the net charge upon the neutron was zero, when the neutron was hit by the neutrino and changed into a proton and an electron, the respective charges of these two particles are positive and negative (the gamma ray has no charge). Hence, when a +1 and a –1 are added, they cancel each other out - making their net charge equal to zero.

Figure 15

Step 1   The anti-neutrino collides with the proton:

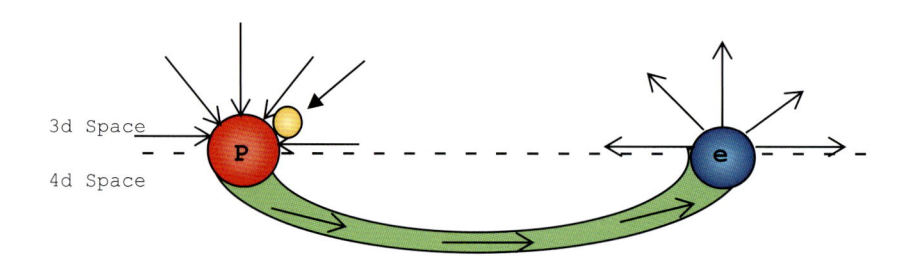

Step 2 The vortex breaks at the proton.

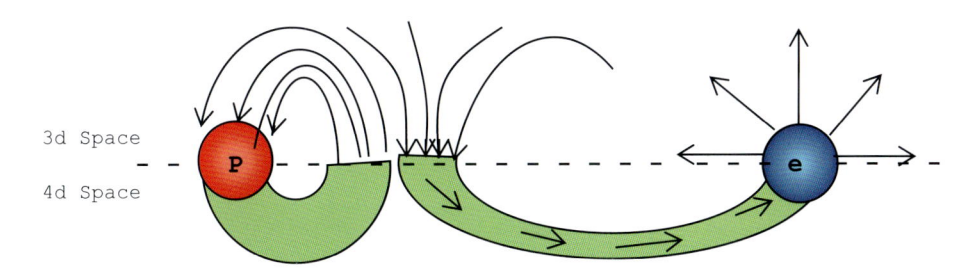

Step 3

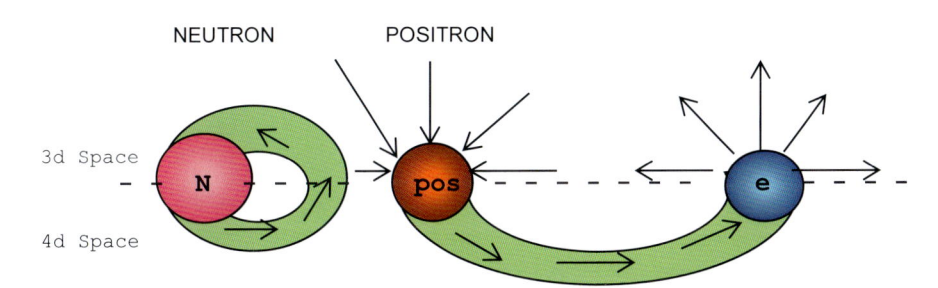

In the last century, the failure to explain the cause of the conservation of charge resulted from a lack of knowledge about the existence of the vortex. But now, it can clearly be seen that these two "particles", (holes in space), are really the two ends of a 4d vortex constructed out of 3d space. Consequently, no matter how many times the vortex is broken, the number of additional entrances and exits into and out of 3d will always mathematically cancel each other out, and their sum will equal the number originally present.

Figure 16

Step 1 The decay of a Neutron:

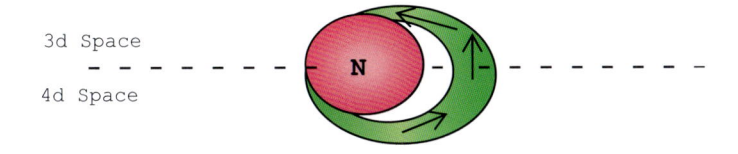

In step 1, notice how the vortex is twisted into a loop. This loop represents a 4d torus that is impossible to draw [it is much like a smoke ring in 3d space].

Step 2 The vortex breaks:

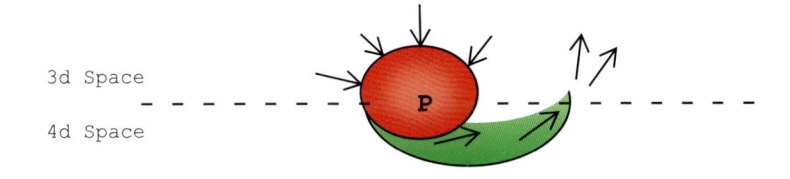

As the vortex breaks, temporarily becoming the "W Particle" in nature, space begins to flow into the proton the vortex surrounded, and begins to flow out of the other end.

Step 3

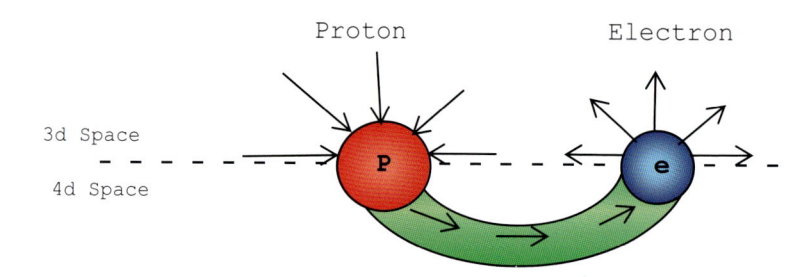

[Diagram showing Proton (P) and Electron (e) connected via 4d Space vortex, with arrows indicating flow in 3d Space]

When the break is completed, the two holes at the ends of the vortex become the proton and the electron.

Another special situation that needs to be addressed is the neutral Kaon. Using this model, when a neutral Kaon decays into two gamma rays, the Kaon's two oppositely charged quarks annihilate; the 3d hole collapses, while at the same instant - the vortex, twisted into a torus, unravels and collapses, and the volume of 3d space that was within it is discharged back into 3d space in opposite directions as gamma rays.

The above illustrations are only some of the annihilations and decays. All of them are beyond the scope of the experiments at this time.

ENTROPY:

Just like the Conservation of Charge, entropy was another observation without an explanation. But a most significant observation! Especially as it relates to the vortices.

According to the Vortex Theory, Entropy is a result of vortices of 3d space flowing from the proton to the electron through 4d space. Consequently, when an atom "absorbs" and then "readmits" a photon, the photon flows into the proton, then into 4d space and through the 4d vortex to the electron; if it is just the right frequency, the electron jumps into a higher "orbit" or energy level. If not, it then flows back out of the electron and into the 3d space from where it came: <u>away from the atom</u>. Hence, because of the 4d vortices, energy is always discharged away from atoms. It does not "build up" within atoms nor create "clumps" of energy as matter does.

Hence, atoms tend to "repel" rather than "collect" energy. Because of this fact, the majority of the photons within any system will eventually be thrown outwards into space, away from the matter they inhabit due to the existence of the vortices: exactly what we have seen in this experiment.

The photons of light were collected by the zinc plate and were transferred to the electrons where they were readmitted back into 3d space. Confirming the above hypothesis, revealing the existence of the vortices via which this "redirection" of energy is taking place.

SUMMATION

The hypothesis that photons of light could be absorbed into atoms through the 3d holes called "protons" and released back into 3d space by the 3d holes we call "electrons" was confirmed. This most elegant proposal was the original purpose and reason for conducting these experiments: to see if light could be transmitted *from* protons in one location to electrons located elsewhere.

The word "*from*" is most important. Since it was theorized that if protons are 3d holes in 3d space that 3d space is flowing into, and if electrons are 3d holes in 3d space that 3d space is flowing out of, then, and most important of all - the flow must be directional. If this hypothesis is true, then, space can only flow from the proton to the electron in 4d space. Unlike quantum entangled electrons or photons that have a volume of whirling space between them [they do not have a volume of space flowing into one hole and out of the other], the journey from the proton to the electron is a one-way trip. To reach the proton from the electron in 4d space, the photon would have to flow in the opposite direction, it would have to go "upstream". Consequently, it was originally theorized that a photon of light can only flow from the proton to the electron in 4d space; it cannot flow from the electron to the proton in 4d space. Amazingly, the sixth experiment, experiment "E" verified this hypothesis.

In summation, it again must be said that perhaps the most amazing observation observed in this experiment is the fact that – according to the physics of the last century – what we see should not be happening at all! However, it is happening.

Because it is happening, even the nonscientist understands that when we see an observation that goes contrary to what we believe we should see, it is the "wake up call" that we need to reexamine our beliefs. Beliefs are merely manmade mental constructs of what we "think" is true. As such, they exist in nature only as the exchange of electrons between neurons in our brains. Most men possess confidence in their constructed beliefs. But when confidence is turned into doubt, the premises upon which the beliefs are formulated must be reevaluated.

CONCLUSION

This experiment reveals the existence of a new subatomic structure in nature: invisible fourth dimensional vortices. These invisible fourth dimensional vortices exist between protons and electrons. Furthermore, these vortices are directional. Because they are directional, they represent flowing space instead of only whirling space.

Being directional, there has to be a reason for their being directional. The reason why they are directional is that their ends are holes in space. Protons and electrons are not "particles" they are only the ends of fourth dimensional vortices. Space is flowing into one hole and out of the other: giving us an elegant explanation for the opposite "charges" of the proton and the electron: allowing us to suddenly, and easily understand a host of seemingly unexplainable phenomena in the universe.

The St. Petersburg State University Experiment That Discovered The "Photon Acceleration Effect"

Konstantin Gridnev, Russell Moon, Victor Vasiliev
Institute of Physics, St. Petersburg State University Russia,
Independent Researcher, U.S.A., Independent Researcher, RUSSIA

ABSTRACT

Using the principles of the Vortex Theory, it was theorized that when a photon encounters an electromagnetic field, both the velocity and the frequency of the photon will change. To prove this revolutionary idea an experiment was devised using a laser interferometer and two electromagnets. The electromagnets were arranged so that when the beam splitter divided the initial beam of laser light into two secondary beams; one of the two secondary beams passed back and forth between the two magnets. With the DC current to the electromagnets turned off, the two beams formed an interference pattern on the target screen. When the current to the electromagnets was suddenly turned on, the pattern fluctuated wildly until the two beams again reached a quiescent state creating a stable pattern on the screen; when the current to the electromagnets was suddenly turned off, again the pattern fluctuated wildly until it reached a quiescent state forming the initial stable pattern on the screen. It was determined that this new effect was a phenomenon created by the changing frequency of the laser light whose velocity is increasing as it passes between the expanding electromagnetic field of the magnets. Because it is a new phenomenon in science revealing that the speed of light is not a constant but indeed can be varied, it possesses great historical significance. It is called the Photon Acceleration Effect.

HISTORICAL BACKGROUND OF THE VORTEX THEORY:

The Vortex Theory is a new, and a brief synopsis of its basic principles is necessary. Just as Einstein's Theory of Relativity is based upon the proposal that time exists as a fourth dimension called space-time and is therefore a fundamental principle of the universe; the Vortex Theory is based upon the proposal that time is not a fundamental principle of the universe. Instead, time exists only as function of motion, a phenomenon created by motion – a shadow of motion. This reduces the five "pieces" of the universe - matter, space, time, energy, and the forces of nature to four: matter, space, energy, and the forces of nature.

Because the construction of each of the four pieces of nature depends upon the construction of the other three, when time is eliminated, eliminating the fourth dimension of space-time; not only does the construction of space have to be re-evaluated, but also, the construction of matter, energy, and the forces of nature.

In an effort to determine how space was constructed when time is eliminated, the idea that space is a void and matter is made of "something" was re-examined. After this study was completed, it was tentatively hypothesized that this relationship was a mistake. That space was made of something and matter was made of nothing; that matter [mesons, leptons, and baryons] exists as three dimensional holes upon the surface of fourth dimensional space; and that space flows into and out of these holes creating the electrostatic force of nature. This hypothesis was given credence when it was further discovered that this new relationship precisely explained the length shrinkage and time dilation effects associated with the Michelson Morley experiment.

For the first time results were submitted in Russia by Prof., Dr. Victor V. Vasiliev at The LIII International Meeting on Nuclear Spectroscopy and Nuclear Structure (NUCLEUS-2003) in Lomonosov Moscow State University. Further results of research are submitted on The LIV International Meeting on Nuclear Spectroscopy and Nuclear Structure (NUCLEUS-2004) in Belgorod State University. The generalization of theoretical researches was made by Mr. Russell G. Moon on The LV National Conference on Nuclear Physics (Frontiers in the Physics of Nucleus) in St.-Petersburg State University.

It must also be noted that this new vision of space and matter is not a return to the old Aether theory. The Aether theory held that both space and matter were made of something. This relationship was much like the relationship between water and ice (the observation that probably inspired this idea).

Energy and the forces of nature are also easily explained by this vision. Energy - photons of light; – are dense regions of space that create their particle effects, while the displacements in space create their wave effects. Although the strong, weak, and gravitational forces are beyond the scope of this paper, the electromagnetic force is explained as currents, or rivers of flowing space. But this explanation seems to create a problem. For if magnetic lines and electrostatic lines of force are created by currents of flowing space, then why isn't matter and energy "caught" up in these currents and carried along with them? For example, when a magnet is held in the air, why doesn't its flowing space move the atoms of air creating a "wind"? Equally important, when a flashlight is held in front of a magnet, why aren't the photons caught in the current created by the electromagnetic field and deflected by the magnetic field?

In an effort to answer these questions, it was easily determined why matter was not deflected [this fascinating discovery will be the subject of its own paper]; but energy was not that easy. However, it was eventually determined that when a photon encountered a region of flowing space, the space [from the photon's point of view] appeared to be denser in the direction from which the space was flowing. This deduction possesses profound consequences.

It possesses profound consequences because it reveals that both the photon's velocity and frequency has to increase. To understand why this increase in velocity and frequency has to occur comes from the Vortex Theory's explanation of gravity.

According to the Vortex Theory, a gravitational field is created by a region of less dense space. This region of less dense space causes the motions of particles and photons to slow down - creating the relativistic effects observed in intense gravitational fields. And, if this is true, when the particles and photons are removed from the less dense region of space and placed in a denser region of space [such as the space between Galaxies], their motions have to speed up. Consequently, when a photon encounters an electromagnetic field, the motion of the space flowing against the photon makes it appear (from the photons point of view), as if the density of the space has increased. This increase in the density of the space flowing against one side of the photon increases the photons velocity in the direction of the field, causing it to turn in the direction of the electromagnetic field. This causes the photons direction of travel to change, making it move at an angle towards the direction of the magnetic field.

| Figure 1 | Figure 2 | Figure 3 | Figure 4 | Figure 5 | Figure 6 |

Photon → Magnetic field

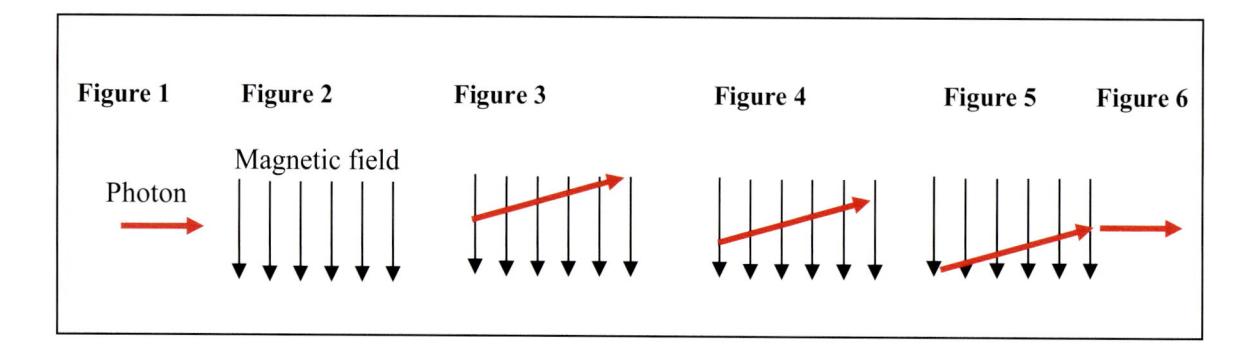

In Figure 1, the photon approaches the magnetic field; in Figure 2, it turns into the field increasing its speed; in Figure 3, the magnetic field pushes down upon it as it tries to move up; in Figure 4, it begins to leave the field; in Figure 5, it resumes its normal speed along its original trajectory as if nothing has happened.

Because the angle of its deflection and velocity ultimately depends upon the strength of the field, the greater the strength of the field, the greater the change in the velocity of the photon and the greater the angle of its deflection; and when this new velocity vector is combined with vector flow of the electromagnetic field, it creates the appearance as if the photon is moving directly through the field without any interference with the field whatsoever!

In the past, this unusual though eloquent effect has made it seem as if the electromagnetic field has no effect upon the photon whatsoever. Furthermore, because of the fact that the photon returns to its original velocity and direction of travel as soon as it leaves the magnetic field, it was believed this effect would be difficult to observe. However, when it became clear that the photon's frequency of vibration also had to change, an experiment was created.

As soon as it became evident that the frequencies of the photons passing through the electromagnetic field were changing, it was easy to devise an experiment. The experiment used an interferometer and two electromagnets.

Because this experiment was treading into "undiscovered territory" - new in the annals of science, it was not known how much of a frequency change could or would be created. Nor could it be predicted how powerful a magnet would be needed. Therefore, in this initial experiment, it was finally decided to merely see if a predicted frequency change could be observed.

To do this, a laser interferometer was constructed. Since it was not known if a pulsed laser would create an undesirable effect, a green Continuous wave laser was chosen having a power rating of 5mW, with a 532 nanometer wavelength. A double concave lens was placed between the laser and the beam splitter. The two first surface mirrors {1&2} were placed 70 cm from the beam splitter; and the electromagnets were positioned 3 cm apart on either side of leg 1 of the laser beam; while the target was placed 50 cm from the beam splitter. The induction of a magnetic field in a flow made 0.3 Tesla.

Figure 7

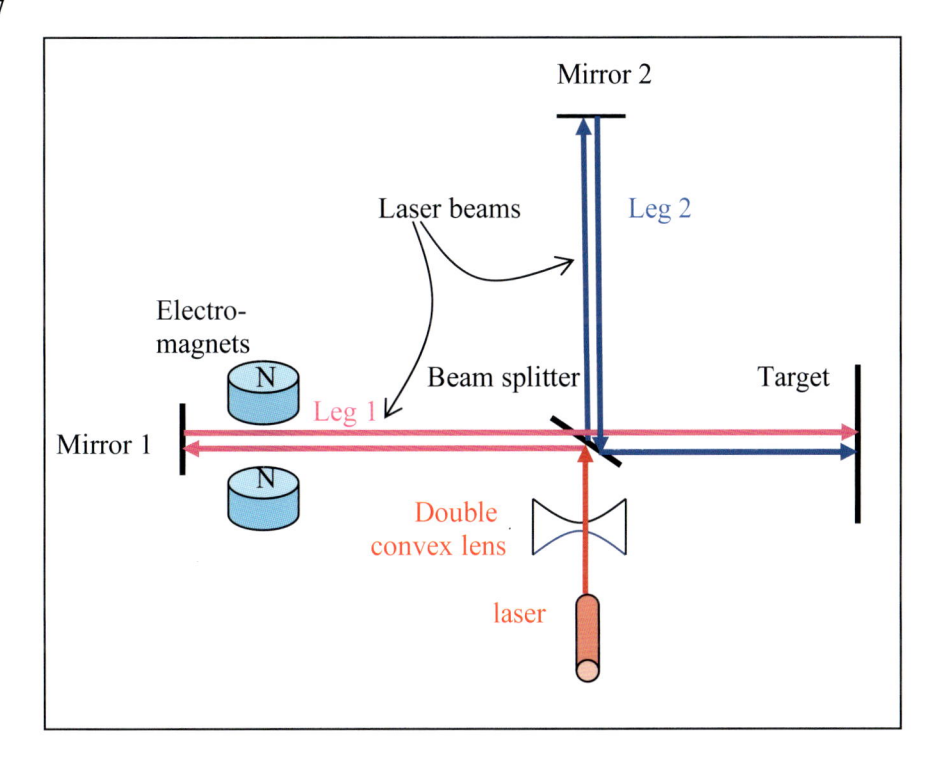

With the electromagnets turned off, the laser was initially turned on. The double convex lens spread out the laser light creating an interference fringe pattern on the target:

Figure 8

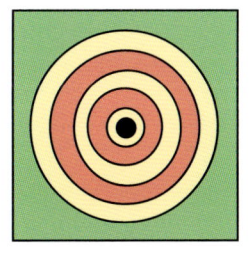

In reality it was wavering slightly. The lines were moving in and out due to vibrations of people walking in the building, minute imperfections in one of the surface mirrors, and temperature variations caused by the heat given off by the individuals witnessing the experiment.

These imperfections forced the witnesses to abandon the room the interferometer was located in and use a television camera and monitor to view the experiment. An air conditioning duct to the room was sealed, the door was shut and the television monitor and power supply for the magnets was set up down the hall. Workmen doing minor repairs to the building were stopped, and foot traffic in the hallways was rerouted.

These changes made the pattern more stable. This stability then allowed the experiment to begin.

PART 1: The experiment began when the current to the magnets was suddenly turned on. And when the magnets were turned on, the pattern on the television monitor fluctuated wildly for several seconds then slowly stopped. When the current was just as suddenly turned off, the pattern again fluctuated wildly and longer than before until it slowly stopped. This procedure was repeated ten times before it began to become evident that the magnets had heated up and were creating air currents that made the pattern unstable.

After approximately 15 minutes, the magnets cooled off and the pattern had again become stable. The entire procedure was repeated a second time with the same results.

PART 2: Later, the experiment was conducted again. This time the current was turned up very slowly. As the current increased, the dark spot in the center turned into a tiny circle and began to expand outward as all of the other circular lines of the pattern increased in diameter too. As the lines increased in diameter, a new dark interference pattern would form in place of the old one and the same sequence of events would continually repeat itself until the current was finally turned all the way up.

When the current was turned back down, the exact reverse would happen. The circular lines of the pattern would continually contract in size until they turned into the dark spot that seemed to continually disappear within the center of the pattern.

THE EXPLANATION OF THE EXPERIMENT

PART 1: The wildly fluttering pattern that was seen on the television monitor appears to be explained by two phenomena occurring simultaneously: {1} the change in the frequency of the photons as the magnetic field increases; and {2} the mutual induction of EMF's from one electromagnet into the other as their magnetic fields expand and contract.

FREQUENCY CHANGE: the change in the frequency of the photons from the laser beam appears to be responsible for part of the observed phenomenon.

When the photons encountered the increasing electromagnetic field of the magnets, according to the principles of the Vortex Theory, their vibrations changed. This continually changed the wavelengths of the photons passing through the fields. Consequently when they re-emerged from their journey between the electromagnets, they were no longer in phase with the photons reflecting off of mirror #2. This caused the interference pattern to change slightly. However, because the electromagnetic field was constantly increasing in strength, the wavelengths of the photons were continually changing as their frequency of vibration changed. But this constant change in the frequency was responsible for only part of the phenomenon observed. The other part was created by the induction of EMFs from one magnet into the other.

THE INDUCTION OF MAGNETIC FIELD'S: the wildly fluttering pattern seen on the screen also appears to be created by the induction of the field from one electromagnet to the other as they are turned "ON and OFF".

When the electromagnets were wired to the DC power supply, the connections were made so as to make one a north south, and the other a south north. This allowed the electromagnetic force to flow directly out of one magnet and into the other. This configuration allowed the photons traveling between them to cross the magnetic field at right angles to the field. However, because the electromagnets are in fact solenoids containing iron cores, a problem is created as they "charge and discharge".

Because of their close proximity, as the DC current to the electromagnets is turned on and the electromagnetic fields begin to expand, this increasing EMF in the first magnet induces an expanding EMF in the second magnet and vice versa. These mutually induced EMF's by each magnet into the other does not allow the core of either magnet to "charge" up to its full electromagnetic strength as it would if the other magnet was not present.

Secondly, when the current is turned off and the electromagnetic fields collapse, the collapsing EMF in one magnet induces an EMF into the core of the opposite magnet. This keeps the EMF of either magnet from collapsing in a linear fashion, causing an oscillating collapse.

PART 2: When the current was increased very slowly, the continuous creation and expansion in the diameter of the lines appears to be caused by a change in the frequency of the photons within the increasing strength of the electromagnetic field.

If the velocity of the photons increased within the increasing strength of the electromagnetic field, the frequency of the photons would have changed. As the photons emerged from the electromagnetic field and returned to their normal velocity, the phase of the photons along leg 1 would be continually increasing. When this continual increase mixed with the unchanging photons of light from leg 2, this continuous increase would be seen in the continuous expansion of the lines of the pattern. Making it appear as if new lines were continually being created from the center of the pattern. There were five fringes created during the increase in the current.

When the current was then turned down very slowly, the strength of the electromagnetic field would decrease and the opposite would happen. The frequency of the photons along leg 1 would be continually changing. When this continual change mixed with the unchanging light from leg 2, this continual change would be seen in the continual decrease in the diameter of the lines of the pattern; making it appear as if the lines were continually disappearing into the center of the pattern. The number of fringes that disappeared were five.

CONCLUSION

Although much more detailed work needs to be done investigating this new and revolutionary phenomenon in nature, it is reasonable to conclude that it was created by a change in the frequency of the photons passing through the electromagnetic field. Also, it must be stated that the discovery of this effect was not discovered by lucky happenstance. This phenomenon was predicted and discovered using the principles of the Vortex Theory. It is tentatively titled: The Photon Acceleration Effect.

EXCITING IMPLICATIONS OF THE EXPERIMENT

The Explanation of Sunspots and solar Flares:

Although the effects of the experiment were artificially created, it appears that this effect also exists naturally in the universe. Amazingly, it suddenly explains one of the curious mysteries of sunspots, and equally important, the creation of solar flares.

Sunspots create a most curious, seemingly contradictory condition: they are cooler than the surrounding surface of the sun; furthermore, most exist for approximately seven days. This means that for seven days, something within their construction is inhibiting the flow of energy into them: an effect that runs contrary to the principles of thermodynamics.

According to the principles of thermodynamics, thermal energy flows from hotter regions into cooler regions. But within the sunspot, something is occurring that is violating this seemingly inviolate law of nature.

Researching the literature about this most curious situation, all sources seem to be in agreement: nobody knows why this phenomenon is occurring. It is one of the great mysteries of the sun. However, the explanation for this phenomenon can be found in the experiment performed at St. Petersburg State University in August of 2005.

When the beam of light passed through the magnetic field, although the speed of light had to increase, only the increase in the frequency of the vibration of photons was observed. But what was not observable was the increase in the speed of light. However, in the sunspot, this increase is what we are in fact observing.

When the sunspot forms, a vortex of gas is created upon the surface of the sun. This vortex extends into the interior of the sun and then re-emerges, returning to the surface of the sun. The spinning ions within the gases would create the intense electromagnetic fields associated with sunspots; making one sunspot the North Pole and the other sunspot the South Pole of the vortex. It is this intense electromagnetic field that is the key to what happens next.

Just as in the experiment, the electromagnetic field within the sunspot causes all photons of light trapped within it to travel at a faster velocity. But not only do they travel at a faster speed within the field, they also move out of the electromagnetic field faster. Hence, the photons, including the infra-red photons, move out of the region within the electromagnetic field faster than the photons from the surrounding regions can move in. Any photons that do move in from the surrounding gas upon the surface of the sun are accelerated, collide faster with other atoms or molecules, are re-admitted faster, and leave the region of the sunspot faster. Consequently, the region of gas within the sunspot is cooler than the rest of the surface of the sun.

This cooler region shows up as a dark spot upon the sun's surface. Equally important is the density of the gas. According to the principles of thermodynamics, the density of the cooler gas has to be greater than the density of hotter gas. And this is what is responsible for the ensuing solar flare.

When the sunspot decays and disappears, according to the literature available, a solar flare apparently occurs. This solar flare would be the result of the sudden influx of thermal radiation from the region of gas immediately surrounding the former sunspot.

When the electromagnetic field decays, the thermal radiation from the region immediately surrounding the former sunspot moves in and is no longer accelerated. Therefore it begins to heat up the cooler gas. As this cooler gas heats up it expands. Its expansion is upward towards the surface of the sun because in this direction there is less pressure. And as it moves upward it continues to expand.

When this expanding long tube of gas reaches the surface of the sun, its velocity hurls it off of the surface of the sun as a long solar flare. Because the sunspot vortex contains charged particles that created its electromagnetic field, these are contained within the gas of the solar flare and are also thrown off of the surface of the sun. It is these charged particles that are attracted by the magnetic north and magnetic south poles of the Earth and are responsible for creating the spectacular Aurora Borealis. But there is also an equally spectacular and previously unknown discovery awaiting us?

This spectacular discovery is a dark spot above both our north and south magnetic poles! Just like the magnetic fields that give the sunspots their ominous dark color, the regions just about the magnetic north and south poles must also possess cooler, darker regions too. However, because the magnetic fields of sunspots are several thousands of times stronger than the magnetic fields of the Earth, and since they create a temperature difference on the surface of the Sun, the dark spots above the magnetic poles of the Earth will only create a two to three degree change in temperature from the surrounding region of air. Consequently, this dark spot will only be visible in the infrared portion of the electromagnetic spectrum. But what a sight it will be.

This sight will become visible via the flyby of a satellite possessing infrared capabilities. The infrared cameras on the satellite must be capable of detecting a temperature change of two to three degrees Fahrenheit. Because earth satellites are now capable of detecting such changes upon the surface of the oceans, this technology appears to already exist. So the only task that remains is to get one to fly over the magnetic north and south poles and take the pictures.

One other interesting effect that appears to be present with these "earth-spots" is the presence of downdrafts. Because the column of air directly above the magnetic north and south poles of the Earth are slightly cooler than the air of the regions surrounding it, this air will sink. This sinking air will create downdrafts that might be visible via the use of a weather radar or Doppler Shift radar upon a satellite. However, to take such a picture, it appears as if clouds will have to be present. On a clear day the satellite will be observing the ground and its temperatures instead of those within the atmosphere.

The Discovery of the Fifth Force in Nature: The Anti-gravity Force

Victor Vasiliev, Russell G. Moon, Marcos Fabian Calvo

All Russian Electrotechnical Institute Moscow, Russia;
Independent Researcher, U.S.A.;
Independent Researcher, Argentina

Abstract

Using the principles of the Vortex Theory, it is theorized that the electron possesses an anti-gravity field. To test this hypothesis, water filled Dissociated Leyden Pipe Capacitors constructed out of long sections of PVC pipe were charged using a Wimshurst Electrostatic Generator. When the inside and outside metal components of the Leyden Jars are removed, and the pipe is stood on end, using a Surface Electrostatic Voltmeter, it is discovered that the charge is not homogeneous; instead, the majority of the electrostatic charge migrated upward towards the top of the pipe, away from the gravitation attraction of the earth.

Keywords: Vortex Theory; Fifth force; Antigravity force; Anti-gravity field; Gravity field; Wimshurst generator

1. Introduction

Without a doubt, the greatest scientific discovery ever made was Isaac Newton's discovery of gravity. No other scientific discovery in the history of mankind has allowed us greater understanding of the universe than the force of gravity. However, it goes without saying that the next even greater scientific discovery is the one that will allow us to defeat gravity: anti-gravity! Anti-gravity engineering combining existing technologies that will allow us, to leave the earth, and travel to the other planets of our solar system as easily and as quickly as we fly today between Europe and America. Anti-gravity engineering will allow this to happen, and is now possible, with this discovery in this paper.

2. Anti-gravity

An in-depth analysis of the Vortex Theory by a team of scientists from Russia, America, and Argentina revealed the possibility of the existence of a fifth force in nature, an anti-gravity force being generated by the electron.

Although when first contemplated, it seems something so revolutionary should have been discovered many years ago; until it was next realized that the existence of an anti-gravity force, could easily have gone completely unnoticed by past investigators. Most of the investigations involving the electron have focused upon its enormous electrostatic charge. Any anti-gravity force would be greatly overwhelmed by the electron's disproportionally large and extremely powerful electrostatic field; making an anti-gravity force almost negligible when the electron is being manipulated by powerful electrostatic or magnetic fields during scientific experiments, or when the electron is part of the atomic structure of an atom and is electro-statically connected to a proton.

However, when the electron is free of the proton, a completely different situation could occur.

When the electron is free of the proton's powerful electrostatic field, and or other electrostatic fields; this anti-gravity force might make its presence known by causing the electron to try to migrate upwards, into the upper atmosphere of our planet. Here, the exotic discovery of "Gigantic Jets" of electrical discharge existing at extremely high altitudes could be the result of large numbers of electrons, attaching themselves to water molecules; moving upward through the atmosphere, and massing at the tops of clouds, until sufficient numbers arrive to create the opportunity for upward discharge.

3. An Experiment

Although the above hypothesis was proposed many years ago, it was not until recently that an experiment was discovered that finally allows us to prove the existence of this anti-gravity force.

This experiment is fairly simple and uses the capacitive principles first discovered in the mid-1740s by Ewald Georg von Kleist; the Leyden Jar, invented by Watson in 1746; and Benjamin Franklin's "Dissectible" Leyden Jar in 1749.

The Leyden jar in this experiment combined a modification of Watson's jar and Franklin's dissectible jar. Its purpose was to trap electrons in the center section of a long vertical tube, set them free, and then see what direction they would move in; and, if they moved at all, would they fill the tube equally, equalizing the charge throughout its volume, or would the majority head towards the top of the tube; creating a disproportionally larger electrostatic voltage at the top vs. that at the bottom of the pipe? For example, if they filled the tube equally, no anti-gravity effect would be noted because the charges at the top and bottom of the pipe would be equal. However, if the majority headed upwards, towards the top of the pipe, creating a difference in the voltage measured at the top and bottom of the pipe, it would indicate that the electrons possessed an anti-gravity field that was propelling them upwards, away from the gravity of the Earth below. To allow the electrons free movement, the Dissectible Leyden Pipe Capacitor seen below in Figure 1 was constructed.

Figure 1

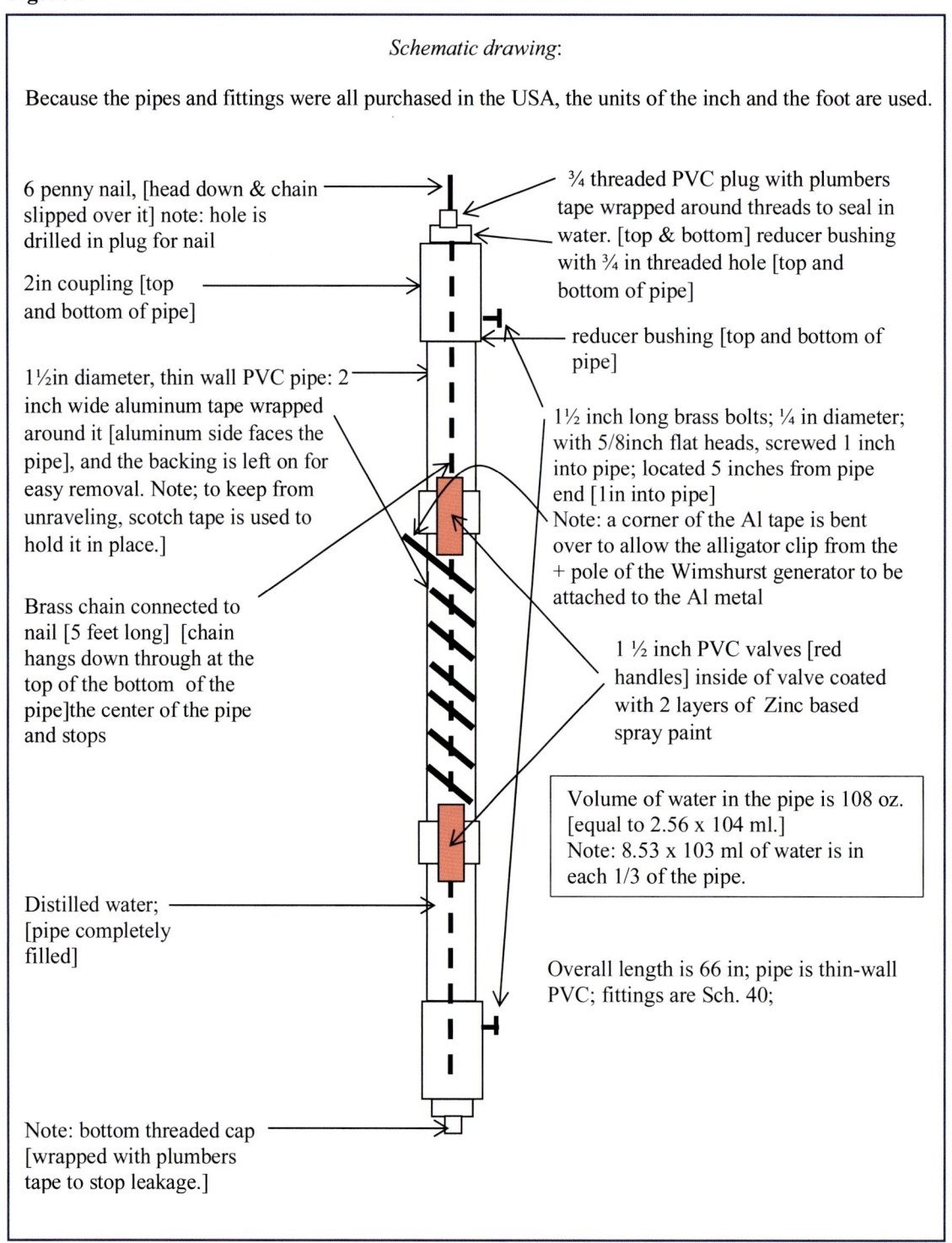

Schematic drawing:

Because the pipes and fittings were all purchased in the USA, the units of the inch and the foot are used.

6 penny nail, [head down & chain slipped over it] note: hole is drilled in plug for nail

2in coupling [top and bottom of pipe]

1½in diameter, thin wall PVC pipe: 2 inch wide aluminum tape wrapped around it [aluminum side faces the pipe], and the backing is left on for easy removal. Note; to keep from unraveling, scotch tape is used to hold it in place.]

Brass chain connected to nail [5 feet long] [chain hangs down through at the top of the bottom of the pipe]the center of the pipe and stops

Distilled water; [pipe completely filled]

Note: bottom threaded cap [wrapped with plumbers tape to stop leakage.]

¾ threaded PVC plug with plumbers tape wrapped around threads to seal in water. [top & bottom] reducer bushing with ¾ in threaded hole [top and bottom of pipe]

reducer bushing [top and bottom of pipe]

1½ inch long brass bolts; ¼ in diameter; with 5/8inch flat heads, screwed 1 inch into pipe; located 5 inches from pipe end [1in into pipe]
Note: a corner of the Al tape is bent over to allow the alligator clip from the + pole of the Wimshurst generator to be attached to the Al metal

1 ½ inch PVC valves [red handles] inside of valve coated with 2 layers of Zinc based spray paint

Volume of water in the pipe is 108 oz. [equal to 2.56 x 104 ml.]
Note: 8.53 x 103 ml of water is in each 1/3 of the pipe.

Overall length is 66 in; pipe is thin-wall PVC; fittings are Sch. 40;

4. Charging and Setup Procedure

To charge the capacitor, the negative side of a Wimshurst machine is attached to the nail using a #8 jumper wire with alligator clips on both ends [negative side of the Wimshurst machine can be determined by turning the hand crank and using the Surface DC Voltmeter to see which side is negative]. Then the positive side of the Wimshurst machine is attached to the aluminum tape using another #8 jumper wire with alligator clips also attached to each end.

Next, the hand crank on the Wimshurst machine is slowly turned 20 times, inducing a static charge of approximately 1.3×10^{-3} coulombs per turn. Then, using thick rubber gloves, the alligator clip is first removed from the nail, and then the other alligator clip is removed from the aluminum tape. Afterwards, and being extremely careful not to touch the nail, the top PVC plug is removed: note, as the plug is removed and pulled away from the pipe, the chain attached to it is then drawn out of the pipe [allowing the aluminum foil to continue to hold the electrons against the inside of the pipe].

Immediately, another ¾ in PVC threaded plug, [with no hole in it], and with 5 wraps of Teflon plumbers tape on its threads is tightly screwed back into the top hole to seal shut the inside of the pipe. Afterwards, the aluminum tape is removed from the outside of the pipe, setting free the electrons from the center 1/3 of the pipe.

Finally the pipe which has always been kept in a vertical position is set upon a piece of wood and leaned up against a wooden ladder or bookcase [no metal touches pipe]. Next, and while wearing rubber gloves, after waiting for ½ hour to allow any currents in the water to subside, and to allow the electrons to move in whatever direction they want to, the two valves are closed; trapping the electrons in the three different sections of the pipe.

Then after waiting for an additional ½ hour, the Surface DC Voltmeter from *Alpha Labs* is turned on and held .5cm from the top bolt and voltage noted; next, it is reset and done likewise to the bottom bolt noting the voltage.

Measurements are then taken every 15 minutes. The first measurement measured the voltage at the top brass bolt, and then the voltage at the bottom brass bolt was measured. After another 15 minutes the voltage at the bottom bolt was measured first, then the voltage at the top bolt. [Note: in tables #1 & #2, the first measurement was made starting at the top bolt, and then 15 minutes later, the next measurement was made starting at the bottom bolt: altering back and forth throughout the test.]

Early on, because it was discovered during testing that the touching of the surface dc voltmeter to the screws or bolts would create huge readings over -30kv that pegged the meter, causing it to be ineffective, the surface dc voltmeter was kept away from the screws and bolts in later tests. Instead of touching the screw or bolt, the brass plate on the back side of the surface dc voltmeter was held .5cm from the screw or bolt on the capacitor. Although this gives a much lower reading, and does not reveal the true amount of charge within the pipe, it does reveal the proportion of the charge now existing within the top and bottom of the pipe.

In test #50, whose graphs [1 & 2] are listed in this paper, the measured charge appears small, [in the hundreds of volts]; however, it must be remembered that if the bolts themselves were touched by the surface dc voltmeter, the readings would be well over -30kV and could not be read by the meter that pegs out at -30kV.

5. Number and Variations of Experiments

It should be noted that fifty different experiments each lasting 6 to 16 hours in duration were performed. A number of different lengths of pipe were used: [18in, 72in, 66in, 70in, 180in]. Several different diameters of pipes were tried: 1in, 1½ in, 2in, 4 in. Different sized screws and bolts were tried: 1/8 wide by 3/4 in long stainless flathead steel screws - installed flush with pipe; 1½ in brass bolts with 5/8 in. diameter flat heads; and 3/8 in diameter by 1 in long stainless steel bolts with rounded 5/8 in diameter heads.

Many different water types were also tried, [tap water; filtered water; distilled water, de-ionized water; HPCL water]. Different chemical solutions were tried: filtered water with Cl_2 added; filtered water with NaCl added; filtered water with HCl added; distilled water with NaCl added; distilled water with HCl added; distilled water with Cl_2 added].

Also, during the tests many different electrostatic charges were used. Six to twenty turns of the Wimshurst machine were tried; with twenty seeming to give the best results.

During initial testing, it was discovered that measuring the charge at the bottom of the pipe could affect the measurement of the top of the pipe and vice versa, so PVC valves were installed. Later, because it was discovered that some of the electrostatic charges within the pipe traveled through the closed PVC valves, the insides of the valves were coated with two thick coats of Zinc based paint: *Zinc Rich Cold Galv, Primer* from the *Sprayon Corporation* [SPR00740].

It is also important to report that the Wimshurst machine can unexpectedly and unpredictably reverse polarity; causing a positive charge to be placed on the chain and a negative charge to be placed upon the aluminum tape. To guard against this unfortunate phenomenon, the Wimshurst machine must be constantly monitored with the DC surface voltmeter to make sure it is putting the correct charge onto the correct elements of the Dissociated Leyden Capacitor.

Finally, it should be mentioned that to reduce static charge upon the technician, the tests were made while standing upon plywood, and an electrostatic hand strap was also used to eliminate static charges on the technician's body. Also, using the hand strap, the body was grounded before each measurement.

6. The Data as seen below in Graphs 1 & 2

In Graphs 1 & 2 below, [test #50], only distilled water was used. Also, each turn of the crank on the Wimshurst machine generated a 60^kV static spark when its two metal end spheres are one inch apart: [given by the manufacturer of the machine *Carolina Biological Supply Co.; N.C., USA*].

Given:

Since: (1) $$F = \frac{C}{V}$$

Where: F = Farads; V = Volts; C = Coulombs

Then: (2) $$C = F \times V$$

Measuring the capacitance of Wimshurst generator's two discharging spheres (located one inch apart): and making the measurement using a Cypress 60-JM117 capacitance meter: a value of 0.11 on the 20nF scale was obtained. This equates to a value of:

$$F = [20. \times 10^{-9}][0.11] = 2.2 \times 10^{-9} = \text{Farads}$$

Since: $V = 60,000 \text{ volts}$

From: (2) $\quad C = F \times V$

Then: $\quad C = [2.2 \times 10^{-9} \text{ F}][6.0 \times 10^{4} \text{ V}] = 1.3 \times 10^{-4}$ coulombs of charge generated per turn of the crank handle on the Wimshurst machine.

Hence, for 20 turns of the handle of the Wimshurst machine in test #31 below, a charge of approximately 2.6×10^{-3} Coulombs was transferred into the Dissociated Leyden Pipe Capacitor: [NOTE: although the amount of static charge transferred into the Dissociated Leyden Pipe Capacitor is fairly important; it is not as important as the movement of the charge inside the pipe after the capacitor is disassembled: see DISCUSSION.]

7. Graph 1

In Graph 1, the long red lines dramatically represent the much larger negative voltage measured on the top bolt vs. the smaller negative voltage represented by the shorter blue lines on the bottom bolt; [with the surface dc voltmeter held at the same distance from the bolt as in the top measurement]. Again, it must be emphasized that the true voltage within the pipe is so large, that the meter could not read it.

8. Graph 2

Graph 2 was added to illustrate the great difference between the two negative voltages at the top and the bottom of the pipe using a line graph representation. It is easy to see the top negative voltage is greater than the bottom negative voltage: revealing that there are more electrons at the top of the pipe than the bottom.

Graph 1 [Test #50]

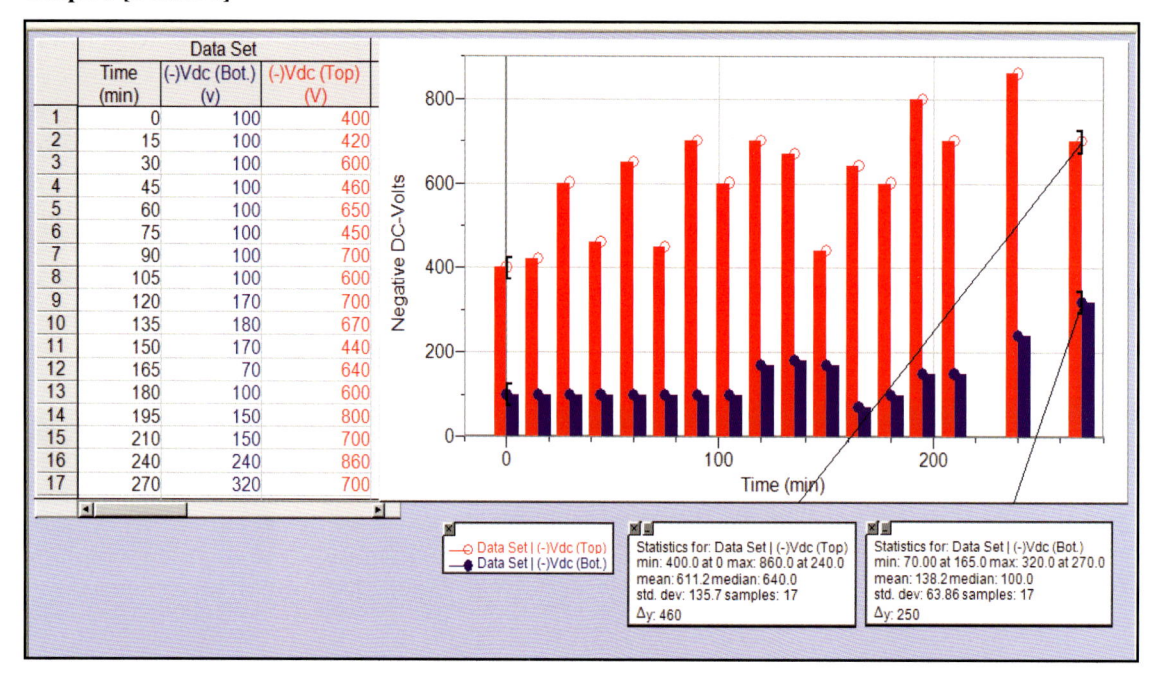

#	Time (min)	(-)Vdc (Bot.) (v)	(-)Vdc (Top) (V)
1	0	100	400
2	15	100	420
3	30	100	600
4	45	100	460
5	60	100	650
6	75	100	450
7	90	100	700
8	105	100	600
9	120	170	700
10	135	180	670
11	150	170	440
12	165	70	640
13	180	100	600
14	195	150	800
15	210	150	700
16	240	240	860
17	270	320	700

Statistics for: Data Set | (-)Vdc (Top)
min: 400.0 at 0 max: 860.0 at 240.0
mean: 611.2 median: 640.0
std. dev: 135.7 samples: 17
Δy: 460

Statistics for: Data Set | (-)Vdc (Bot.)
min: 70.0 at 165.0 max: 320.0 at 270.0
mean: 138.2 median: 100.0
std. dev: 63.86 samples: 17
Δy: 250

Graph 2 [Test #50]

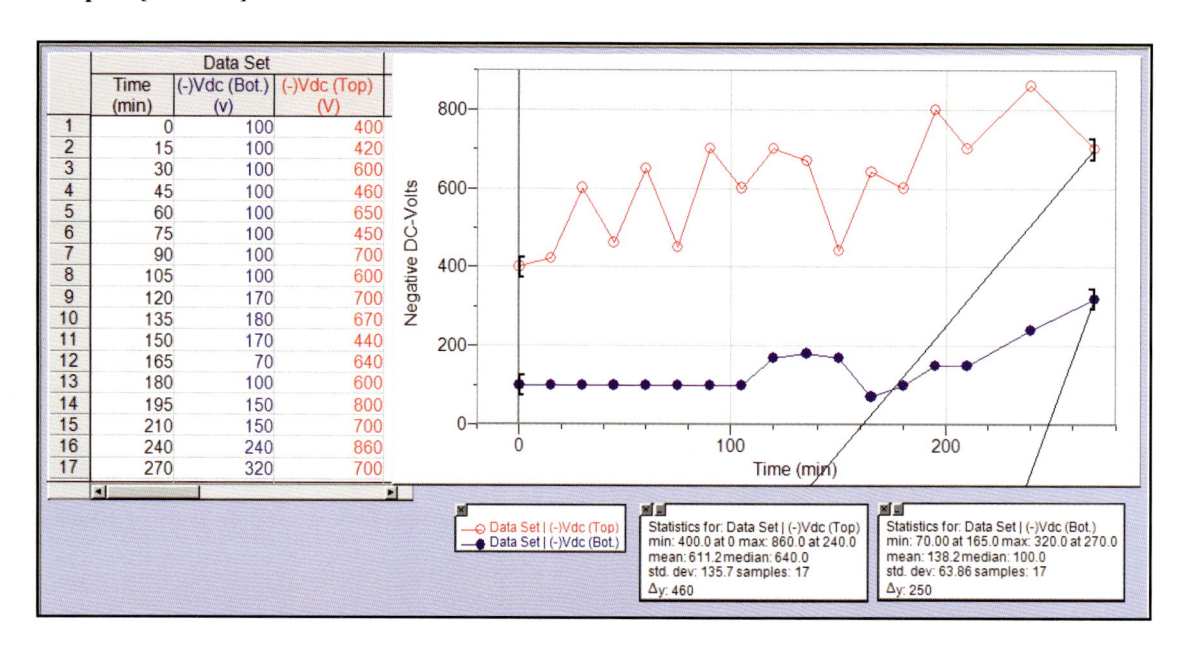

#	Time (min)	(-)Vdc (Bot.) (v)	(-)Vdc (Top) (V)
1	0	100	400
2	15	100	420
3	30	100	600
4	45	100	460
5	60	100	650
6	75	100	450
7	90	100	700
8	105	100	600
9	120	170	700
10	135	180	670
11	150	170	440
12	165	70	640
13	180	100	600
14	195	150	800
15	210	150	700
16	240	240	860
17	270	320	700

Statistics for: Data Set | (-)Vdc (Top)
min: 400.0 at 0 max: 860.0 at 240.0
mean: 611.2 median: 640.0
std. dev: 135.7 samples: 17
Δy: 460

Statistics for: Data Set | (-)Vdc (Bot.)
min: 70.0 at 165.0 max: 320.0 at 270.0
mean: 138.2 median: 100.0
std. dev: 63.86 samples: 17
Δy: 250

9. Analysis of Test #50

Test #50 took approximately 6 hours to complete. This test followed the protocol as explained in the CHARGING AND SETUP PROCEDURE section of this paper. Humidity 44%; air temp 81^0; distilled water [supplied by Zephyrhills], temp 80^0; water resistance 200kΩ; Ph 4.4. Measurements were initially taken every 15 minutes. It was observed that the reading at the top of the pipe always exceeded the bottom readings. And even though the readings at the bottom of the pipe began to increase after a few hours, they never came close to equaling the much higher readings taken at the top of the pipe.

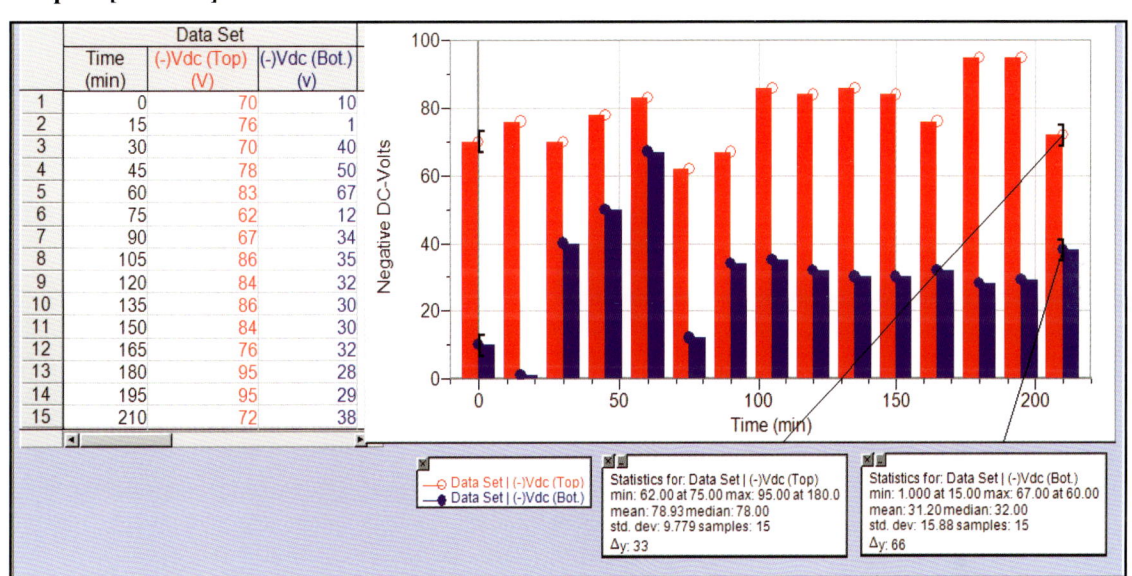

Graph 3 [Test #45]

#	Time (min)	(-)Vdc (Top) (V)	(-)Vdc (Bot.) (v)
1	0	70	10
2	15	76	1
3	30	70	40
4	45	78	50
5	60	83	67
6	75	62	12
7	90	67	34
8	105	86	35
9	120	84	32
10	135	86	30
11	150	84	30
12	165	76	32
13	180	95	28
14	195	95	29
15	210	72	38

Statistics for: Data Set | (-)Vdc (Top)
min: 62.00 at 75.00 max: 95.00 at 180.0
mean: 78.93 median: 78.00
std. dev: 9.779 samples: 15
Δy: 33

Statistics for: Data Set | (-)Vdc (Bot.)
min: 1.000 at 15.00 max: 67.00 at 60.00
mean: 31.20 median: 32.00
std. dev: 15.88 samples: 15
Δy: 66

10. Test #45, Eliminating Unforeseen Variables: the Chain

Tests #36-45 were undertaken to determine if the removal of the chain from out of the top of the pipe could in any way be causing the majority of the electrons to congregate at the top. To eliminate any affects of the chain, immediately after charging the pipe, removing the chain, and recapping it, the pipe was turned upside down; then aluminum tape was removed, and the pipe was allowed to set for ½ hour before closing the two valves; and then as before, waiting for another ½ hour before beginning to take measurements.

The results of Test #45 can be seen in Graphs 3 & 4. Again, note how the voltage measured at the top of the pipe [red lines] are larger then the voltages taken at the bottom of the pipe [blue lines]; indicating that the charge at the top of the pipe is greater than the bottom; revealing that the majority of the electrons have migrated to the top of the pipe. Consequently the removal of the chain had no affect on the movement of the electrons upwards, towards the top of the pipe.

Graph 4 [Test #45]

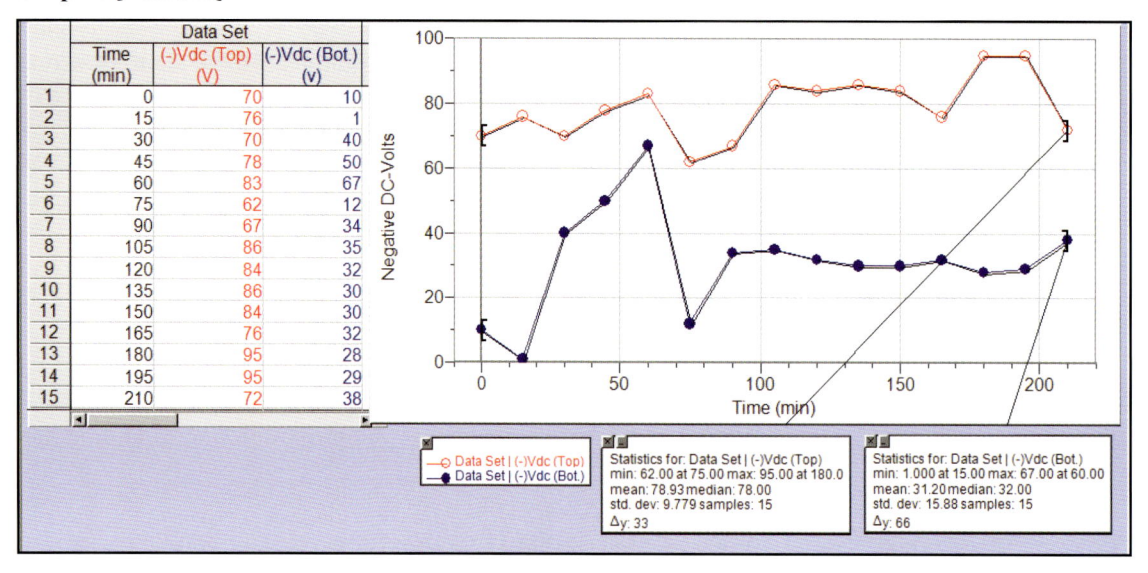

#	Time (min)	(-)Vdc (Top) (V)	(-)Vdc (Bot.) (v)
1	0	70	10
2	15	76	1
3	30	70	40
4	45	78	50
5	60	83	67
6	75	62	12
7	90	67	34
8	105	86	35
9	120	84	32
10	135	86	30
11	150	84	30
12	165	76	32
13	180	95	28
14	195	95	29
15	210	72	38

Statistics for: Data Set | (-)Vdc (Top)
min: 62.00 at 75.00 max: 95.00 at 180.0
mean: 78.93 median: 78.00
std. dev: 9.779 samples: 15
Δy: 33

Statistics for: Data Set | (-)Vdc (Bot.)
min: 1.000 at 15.00 max: 67.00 at 60.00
mean: 31.20 median: 32.00
std. dev: 15.88 samples: 15
Δy: 66

Because it was found that much larger interior negative charges cause the outside of the pipe to become positive, making it harder for the charge to move about freely when the pipe is turned upside down; for Test #45, a much lower charge was induced into the Dissociated Leyden Pipe Capacitor by using only 6 turns of the Wimshurst generator: [1.3 x 10^{-4} coulombs / (per turn of the crank)] x 6 turns = 7.8 x 10^{-4} coulombs].

11. Eliminating Unforeseen Variables: the lingering internal charge in the capacitor

During these tests, a strange phenomenon was observed. It was discovered that re-assembling the disassociated capacitor and trying to discharge it cannot discharge the upper 1/3 and lower 1/3 regions of the PVC pipe: which now have a negative residual charge resulting from electrons being imbedded in the plastic, especially in the top 1/3; [one top reading was over -30kV]. In fact, after reassembly and refilling the capacitor with water, but before charging; touching the meter to top bolt, actually pegged the meter.

Consequently, through trial and error, it was discovered that carefully rinsing and re-rinsing the pipe using a 50/50 mixture of water and HCl for at least .5 hr discharged it. [If not done, the lingering internal negative gives misleading results in future tests]. However, for a sequence of 3 tests using high charges; after rinsing, and reassembling the capacitor, pretesting the charges at the bolts revealed the lingering charge did not disappear; and the cleansing of the pipe had to be done all over again.

Unfortunately, it was also found that when the residual charge was too high, and several cleansing were unsuccessful, the pipe had to be abandoned and a completely new capacitor had to be constructed.

It was also discovered that the act of removing a length of PVC pipe from a rack of pipes by sliding it out, caused it to pick up static charge. This charge was still present before the apparatus was charged with the Wimshurst machine and had to be eliminated by again re-rinsing the capacitor with the 50/50 solution. Later, to eliminate any charge, new pipes and fittings were pre-washed with the solution of HCl and water before being glued.

Also, it was found that the experiment did not work well with humidity readings above 48%. Hence, a dehumidifier was used to control the humidity.

12. Eliminating Unforeseen Variables: the Surface dc Voltmeter

To eliminate any affects created by the surface dc voltmeter during the measurement of the static voltages within the pipe, Test #45 was made by holding the Surface dc Voltmeter .5 cm. from the bolts on the pipe. Although this results in a much lower voltage than the touching of the meter to the bolts, it lessens the ability of the meter to move the charges within the pipe.

Next, and again, to eliminate the ability of the surface dc voltmeter to move the electrostatic charges within the pipe, the sequence of measuring voltages at the top and the bottom of the pipe were altered:

The first sequence was started by first measuring the voltage at the top of the pipe, and then measuring the voltage at the bottom of the pipe; then, for the next sequence 15 min. later, the bottom voltage was measured first, and then the top voltage was measured afterwards. This alternating of the sequence back and forth was continued throughout the experiment.

13. Discussion of Results

It is important to remember that in all 50 tests, the aluminum foil was wrapped only around the middle one third of the pipe; this is where the positive pole of the Wimshurst machine was attached to via a jumper cable with alligator clips. Consequently this was the location where the negative charges of the electron's being fed into the interior of the Dissociated Leyden Pipe Capacitor were initially attracted. However, when the chain and the foil were removed, the electrons were suddenly set free and were able to move freely about the interior of the pipe.

Logic tells us that the electrostatic charges of the electrons should force them to spread out equally throughout the volume of the pipe: creating the same electrostatic dc voltage readings at both the top and bottom of the pipe. But this is not what happened.

Instead of creating a homogeneous volume of electrostatic charge equal throughout the pipe, the majority of the electrons moved upward and concentrated towards the top of the pipe. And, after the valves were closed, and electrostatic measurements began, it was easily seen that the charge trapped in the very top of the pipe is much greater than the charge trapped in the very bottom of the pipe.

Also, although it might be speculated that unknown, more massive positive ions [with greater specific gravities] in the water collected the electrons, were neutralized and sank allowing less massive, negatively charged ions [of lesser specific gravities] to rise, creating the different readings at the top and bottom of the pipe; this argument cannot account for the residual negative charge attached to the bottom and top thirds of the pipe after the test is over and the water with ions is removed.

14. In Conclusion

The unusual upward motion of the electrons in this experiment forces us to conclude the following: that the electron possesses an antigravity field; and as theorized at the beginning of this paper, there is a fifth force in nature: an anti-gravity force, that is in direct opposition to the force of gravity!

THE END?
ABSOLUTELY NOT!
IT'S THE BEGINNING OF A BRAND NEW ERA IN THE HISTORY MANKIND!!!